Quantitative Methods in the Humanities and Social Sciences

Quantitative Methods in the Humanities and Social Sciences is a book series designed to foster research-based conversation with all parts of the university campus – from buildings of ivy-covered stone to technologically savvy walls of glass. Scholarship from international researchers and the esteemed editorial board represents the far-reaching applications of computational analysis, statistical models, computer-based programs, and other quantitative methods. Methods are integrated in a dialogue that is sensitive to the broader context of humanistic study and social science research. Scholars, including among others historians, archaeologists, new media specialists, classicists and linguists, promote this interdisciplinary approach. These texts teach new methodological approaches for contemporary research. Each volume exposes readers to a particular research method. Researchers and students then benefit from exposure to subtleties of the larger project or corpus of work in which the quantitative methods come to fruition.

Edward Beltrami

Digressions in Elementary Probability

The Unexpected in Medicine, Sports, and Society

Springer

Edward Beltrami
Applied Math
SUNY Stony Brook
Setauket, NY, USA

ISSN 2199-0956 ISSN 2199-0964 (electronic)
Quantitative Methods in the Humanities and Social Sciences
ISBN 978-3-031-70050-7 ISBN 978-3-031-70051-4 (eBook)
https://doi.org/10.1007/978-3-031-70051-4

This Springer imprint is published by the registered company Springer Nature Switzerland AG
The registered company address is: Gewerbestrasse 11, 6330 Cham, Switzerland

Preface

This book is an introduction to the mathematics of chance hoping to untangle some of the unexpected events and misleading claims engendered by the intrusion of chance into the public discourse. I want the topics to be relevant to current events and so there are many quotes from newspaper articles (usually *The New York Times*). There are several excellent books that purport to do the same thing, but they totally eschew the math. However, the inclusion of elementary ideas from probability and statistics as we do here gives the discussion more heft and insight. To make an analogy think of a prose translation and commentary of some important work written in a difficult vernacular (Dante's *Commedia* comes to mind). The commentary serves to elucidate the nuances that otherwise would be partly obscure. The mathematics provides a similar role here though, to be truthful, some of it can be partially bypassed without too much of a loss. The real story lies with the examples and the interpretative gloss that goes with them.

In this book, the interplay between chance and order is limited mostly to binary events such as success/failure as they occur in a diversity of interesting applications that I present as a series of vignettes. The goal is to entertain and instruct with topics that range from unexpected encounters with chance in everyday experiences to significant "must know" insights regarding human health and other concerns in the social sciences. There is quite a bit of material in this book that is not readily found in other references, if at all. My hope is that one can gain valuable insight into the workings of happenstance by reading through the individual vignettes some of which are loosely interconnected but most of them are independent digressions. Since the book is expository and not a text, there are no exercises for the reader.

Though the sub-title to this book uses the benign term "unexpected," a number of applications involve more bizarre encounters with randomness in which one's first reaction is that of confronting a paradox. Indeed at one place or another, we meet apparent "paradoxes" that go by various names, each of them instructive and amusing simultaneously.

The first section provides the tools for being able to discuss random sequences with hints at what is to follow. Then I discuss a well-known event in baseball which provides a testing ground for distinguishing between skill and chance. The Poisson distribution plays an essential role here and in the next section in which a classic experiment in genetics is reviewed. Then, in preparation for a subsequent application to basketball, I discuss a curious and somewhat counter-intuitive bias that occurs in certain sequences of Bernoulli trials based on recent important work of Miller and Sanjurjo. This is then applied to the "hot hands" phenomenon in sports. An alternate take on all this then follows leading to the link between restricted choices and Bayes Theorem and the notion of cognitive illusions, a topic of interest in psychology. This is followed by another surprising and, to some extent, bizarre result known as Stein's Paradox and apply it to baseball.

The first seven vignettes have a common outlook with a nod to sports. The next five sections have medical concerns at their core.

The troublesome topic of disease clusters, namely to decide whether the clumping of events is due to chance or some environmental cause, is treated using both the Poisson and normal approximations to the binomial distribution and this leads naturally into a discussion of the base rate fallacy and a case study of hospital performance. Next, another medical case study, this time concerning some tricky questions about the effectiveness of colonoscopy and other medical interventions. This is the most provocative topic in the book because it suggests that colonoscopy may be an over-used diagnostic. I follow this with a discussion of curious fallacies that go by the names of Simpson's paradox and Berkson's paradox. Then, a brief treatment of probabilities and hypothesis testing in clinical trials. An application follows in a reprise of the "hot hands" controversy that was considered earlier, this time with a different approach.

Next considered is the error in random sampling when polling for candidate preference with specific current examples. The essential tool here is covariance of random variables. I follow this with a treatment of the spooky quality of coincidence using appropriate mathematical tools.

After this, code breaking at Bletchley Park during the Second World War using Baye's theorem. Then I return to Poisson events to discuss another unexpected result, followed by the use of spatial Poisson events in the delivery of emergency response services. The last three sections require somewhat more involved mathematical consideration regarding Poisson processes and the notion of entropy. Though the exposition is generally accessible, a reader may prefer to skim these sections on a first reading.

Finally, in the last two vignettes, I give an account of bizarre fluctuations that occur in a run of Bernoulli trials followed by a surprising coin tossing game that conjures Parrondo's Paradox. These serve as bookends to the very first section of the book, but the results now are more counter-intuitive and the last vignette in particular suggests an intriguing possibility for managing a stock portfolio. The probability tool here is that of Markov chains which is introduced without undue technicalities. A more complete but still elementary treatment of such chains is in the first chapter of my book [23] where one finds an application to models of social mobility, a topic that meshes well with the ideas covered in the present work. It is worth mentioning in this regard that some of the topics I treat are given more exhaustive coverage as detailed case studies in a book by Hubert and Wainer [22], and it is warmly recommended as a supplement to what I present.

The probability theory involved is elementary using the binomial theorem and its extensions to Poisson and normal events in addition to conditional probability and covariance. Any unusual probability theory that is needed, stuff not regularly found in standard texts, is included as we go along. As mentioned earlier, it is usually the case that one need not follow the mathematical arguments in detail to gain an understanding of what is going on. The math is there for completeness and because some readers appreciate the added nuance. In any event, I provide an optional brief Tutorial at the end that covers the basic ideas in probability and statistics needed in the main part of the book. In this Tutorial, I also discuss the frequentist versus Bayesian approaches to uncertainty.

Although this book adopts an informal stance in presentation, I retain the more formal habit of labelling a statement as a Theorem when it is followed by a detailed proof.

Besides a list of references, I include an Appendix devoted to several codes written in MATLAB that were used to illustrate various topics in the text as well as to support several figures that appear throughout. The codes are included to provide validation for the curious reader.

Although additional topics of interest could have been included, such as random processes in communication theory, the mathematical tools needed are more advanced and therefore not suitable for this book.

Setauket, NY, USA Edward Beltrami

Contents

Chapter 1
Success Runs in Bernoulli Trials

A streak of 76 consecutive heads occur in a coin tossing game. The loser remarks that *"A weaker man might be moved to re-examine his faith, if in nothing else than in the laws of probability"* (from the play *Rosencrantz and Guilderstern are Dead* [Tom Stoppard, 1967]).

I want to comment on random events that occur sequentially in time or space with probability p of a success at different instants of time or at disjoint portions of space. To be specific there s a random variable X_j that takes on the value one with probability p if the event occurs at the *jth* trial and is zero otherwise. One then commonly seeks the probability of k successes in n trials, namely the sum $\sum X_j$ with j summed over all n trials and this will be supplied by the *binomial distribution*. One of my goals is to look at some of the conundrums that arise in a sequence of binary events labeled as success or failure, as in the heads or tails example, or more generally, as one and zero indicating success and failure.

Examining success runs and their lengths raises the interesting question of how people perceive randomness. The tendency is for individuals to reject patterns such as a long run of successes or failures as not typical of randomness and to compensate for this by judging frequent alternations between zeros and ones to be more typical of chance; a lengthy sequence of consecutive heads or tails, or of any other binary variable, is seen as due to something more than luck. Experiments by psychologists (Bar-Hillel and Wagenaar [1], Falk and Konold [2], and the overview in [3]) who ask subjects to either produce or evaluate a succession of digits reveal a bias in favor of more alternations than an acceptably random string can be expected to have; people tend to regard a clumping of digits as a signature pattern of order when in fact the string is randomly generated.

Closely related to this cognitive flaw is the gambler's fallacy of supposing that after a long run of failures there is a higher probability of success than is warranted by Bernoulli trials. Bar-Hillel and Wagenaar [1] comment that an individual's assessment of randomness in tosses of a fair coin seems to be based on the "equal probability of the two outcomes together with some irregularity in the order of their appearance; these are expected to be manifest not only in the long run, but even in

E. Beltrami, *Digressions in Elementary Probability*, Quantitative Methods in the Humanities and Social Sciences, https://doi.org/10.1007/978-3-031-70051-4_1

relatively short segments-as short as six or seven. The flaws in people's judgments of randomness in the large is the price of their insistence on its manifestation in the small."

Such an observer may conclude that an unusual clump of successes is evidence of a notable event taking place, something not due solely to chance. It seems that we are good at detecting patterns and attaching significance to them. This is especially true in activities carried out by people such as fund managers or athletes where there appears to be a tradeoff between luck (chance) and skill; I will try to shed some light on this with a detailed look later at the remarkable hitting streak of Joe Di Maggio in his anno mirabilis of 1941.

After this I inject a moment of cognitive dissonance into coin tossing based on a remarkable insight of Miller and Sanjurjo [5] in which they reveal a subtle bias that crops up when attempting to compute the probability that a head follows another head in a sequence of independent tosses. Psychologists Tversky and Gilovich attribute the perception of "hot hands" among fans and players of basketball to be a consequence of thinking that a run of successes in a short string of throws must be due to more than chance. I quote from their paper [7]: "The belief in the hot hand is merely one manifestation of this fundamental misconception of the laws of chance. Maybe the streaks of consecutive hits that lead players and fans to believe in the hot hand do not exceed, in length or frequency, those expected in any random sequence." But, as will be seen, there is more to this story, doubts sown by the unexpected statistical bias I mentioned above which suggests that this conclusion may be premature.

This is a good place to include a quote from a NY times article by Richards [69]: "We are in the habit of looking in the past in the hope of seeing a pattern and then projecting that pattern into the future. But when we pull from small a small sample size we can end ups guessing wrong... when we look for a pattern the odds are that we'll find one."

There is an oft-told illustration of how people misconstrue success runs in coin tossing. On the first day of a probability course the instructor asks his students to record 200 flips of a fair coin or, instead, they may choose to pretend to flip and fake the results. The next day he amazes them by glancing at each of the student's papers and reporting, in nearly all cases, which are true coin tosses from faked data. His ability to do so is based on the surprising fact that in a sequence of 200 tosses it is extremely likely that a run of six or more consecutive heads or tails will occur, as I show below. However, the typical person will rarely include runs of that length.

To examine this more closely I need to establish a formula due to Berresford [12] for the probability of success runs of a given size. A proof is given below and a MATLAB code (*Berresford.m*) to implement the algorithm will be found in the Appendix.

Theorem 1 The probability P(n, k) of a success run of length k or more in n Bernoulli trials with probability of success p is given by

$$P(n,k) = P(n-1,k) + (1 - P(n-(n+k)),k)\,p^k(1-p)$$

Proof Define a run, clump, or streak, of size k in a string of n coin tosses as a sequence of exactly k successive heads. We want to compute the probability P(n, k) of finding runs of size ≥k. To do this we consider two mutually exclusive events:

(i) There are runs of size ≥k among the first n – 1 tosses.
(ii) There are no runs of size ≥k in the first n – 1 tosses but the last k tosses out of n do form a clump.

Note that (ii) is the intersection of two independent events, the first being that the last k + 1 tosses are of the form tail followed by k heads (if these k + 1 tosses were all heads there would be a run of size ≥k among the first n – 1 tosses, which we exclude), and the second event is that there is no run within the first n – k – 1 tosses.

The probability of (i) is P(n – 1, k) whereas ii has the probability (1 – P(n – k – 1, k)) times $p^k(1 - p)$. Because (i) and (ii) are disjoint,

$$P(n,k) = P(n–1,k) + (1–P(n–(n+k)),k)p^k(1–p)$$

To begin the recursion one has a couple of immediate identities: P(n, k) = 0 for all n < k, and P(k, k) = p^k since a k-clump occurs in only one way in k tosses. Note that all runs with clumps greater or equal to k is the disjoint sum of all clumps of size greater or equal to k + 1 plus all clumps whose size is exactly k. It follows, therefore, that the probability of a run of length *exactly k* is

$$P(n,k) - P(n,k+1)$$

A slight modification of the argument that leads to Theorem 1 gives us the probability Q(n, k) of having runs of length greater than or equal to k of either heads or tails (Berresford [12]). I limit myself this time to p = 1/2. Once again there are two disjoint events. Either, there is such run among the first n – 1 tosses or ii, there is no such run among the first n – 1 tosses but there is one that involves the last toss. The first event has probability Q(n – 1, k) while ii is the product of two independent events, the first being that there is a run of k heads or tails in the last k tosses and the second event is that there is no such run among the first n – k tosses. As Berresford shows this can happen in one of two ways: the last k tosses are heads preceded by a tail or the final k tosses are tails preceded by a head. This means that there are no runs of length ≥k among the first n – k tosses that end in tail in one case or that end in head in the alternate case. This results in

$$Q(n,\ k) = Q(n-1,\ k) + (1 - Q(n-k,k))/2^k$$

with initial conditions Q(n, k) = 0 for n < k and Q(k, k) = $1/2^{k+1}$

Next, let's use Theorem 1 to compute P(200, 6) and give an answer to the question posed earlier, the one in which the instructor asked his students to toss a coin 200 times. The likelihood that there is a clump of heads of size ≥6 is .8009, a better than 80% chance and there is a better than even chance, .5437, of a head run of length ≥7. These unexpected results are what confounded the students.

To put the formulas in to their simplest context I list here the sixteen strings associated with n = 4:

1 1 1 1	1 0 1 1	0 1 1 1	0 0 1 1
1 1 1 0	1 0 0 1	0 1 1 0	0 0 1 0
1 1 0 1	1 0 1 0	0 1 0 1	0 0 0 1
1 1 0 0	1 0 0 0	0 1 0 0	0 0 0 0

From Theorem 1 we find that P(4, 2) = .5 and P(4, 3) = .1875 and these values are immediately verifiable from the above table since there are eight runs of size ≥2, giving a probability of 8/16 = .5, and three runs of size ≥3, leading to a probability 3/16 = .1875

Let me mention, as an afterthought, that in computing the probability of a clump of exactly k heads in n tosses of a fair coin one might be tempted to use the following argument. Let the first k tosses be heads, succeeded by a tail. Next, slide this k-clump to the right one place and bookend it with tails. Continue with this sliding procedure until you reach the last k positions, all heads, preceded by a tail. Each of these events has probability $\frac{1}{2}^{k+2}$ except for the first and last clumps which have probability $\frac{1}{2}^{k+1}$. The sum of these probabilities is then, or so it would seem, the desired final probability. But this would be incorrect. To see this consider a string of length 8 and suppose k = 3. Let * indicate either a head or tail. Then we have six events

HHHT****
THHHT***
THHHT**
**THHHT*
***THHHT
****THHH

The $2^4 = 16$ possibilities for the first event includes the string HHHTHHHT and this is also one of the $2^3 = 8$ possibilities for the fifth event (there are others). Since these events overlap we cannot add their respective probabilities of $1/2^4$ and $1/2^5$ since otherwise we get an over-estimate for the final probability of a clump of exactly size k.

I mentioned that a random binary sequence is expected to have runs of a length longer than what would be anticipated in an invented sequence. There is a caveat however. A calculation carried out in a paper by Bloom [11], Theorem 2 below, establishes that on average about a quarter of all tosses of a fair coin are alternating head-tail or tail-head flips. A sequence with too many singles tends to have a smaller number of gaps available for long runs of either heads or tails so does the failure of true randomness reside in a scarcity of long success runs or a paucity of singles ? I investigated this conundrum by generating 25 random binary strings of length

$2^8 = 256$ and found that both Lemma 2 and 3 are nicely vindicated showing that a quarter of tosses can be singletons without compromising the occurrence of long streaks of heads and tails. This is empirical evidence that singletons and long gaps are compatible; having too many of one will dilute the other, however, and this would cast some doubt on whether the sequence is generated by a fair coin.

Theorem 2 Let X_n be the number of singles in n tosses of a fair coin for n no less than 2. Let S_n be the number of sequences on n tosses containing exactly k singles. The expected number of singles in n tosses is (n + 2)/4.

Proof We can count the number of times a flip is a single by letting $Z_i = 1$ if the ith toss is a single and zero otherwise. Then $X_n = Z_1 + \ldots + Z_n$. Now $Z_i = 1$ iff both $Zi \neq Z_{i-1}$ for $i > 1$ and $Z_i \neq Z_{i+1}$ for $i < n$. The expected value of Z_i, or, equivalently, $\Pr(Z_i = 1)$, is ½ if i = 1 or n and $1/2^2 = 1/4$ if $1 < I < n$.

Therefore the expected value of X_n is $1/2(2) + 1/4(n - 2) = (n + 2)/4$.

We can check the result of Theorem 2 in the case of n = 4 by looking back at the 16 possible sequences that I gave earlier. One finds 4 sequences with one singleton, 6 sequences with 2 singletons, and 2 more sequences with 4 singletons giving a mean value of 3/2 which is precisely (n + 2)/4.

A slight modification of the proof gives the expected number of singleton heads in n tosses. It suffices to notice that $Z_i = 1/2$ instead of 1 in the above proof giving an expected value of (n + 2)/8 in this case. Again, looking at the table of 16 sequences used above, we now find that there are 6 sequences with one head singleton and 3 sequences with 2 head singletons for an average of 3/4 as expected.

I close with another fact about runs, a proof of the expected number of runs of heads in any randomly chosen sequence of n tosses and m heads (taken from the book by Ross [13]). This result is related to ideas discussed later regarding a subtle bias.

Theorem 3 The expected number of head runs in n tosses of a fair coin in which m denotes the number of heads is (m/n)(n – m + 1).

Proof Let I_j be the indicator variable which equals one if the *jth* toss is the beginning of a head run and zero otherwise. The expected number of runs is then $\Sigma E(I_j)$ summed from 1 to n. The probability that there is a head in position 1 is $E(I_1) = m/n$ and, for $j = 2, \ldots, n$, the probability that there is a T in position $j - 1$ followed by a H in position j is

$$E(I_j) = m/n(n - m)/(n - 1).\text{Therefore}$$

$$\Sigma E(I_j) = m/n + (n - 1)[m/n(n - m)/(n - 1)] = (m/n)(n - m + 1).$$

A similar expectation for the number of tail runs is exactly the same except that m is replaced by n – m. It then follows that the expected number of runs of either H or T is 1 + 2m(n–m)/n

In Tom Stoppard's *Rosencrantz & Guildenstern are Dead* that was quoted in the beginning of this section we can also read "*the equanimity of your average tosser of coins depends on a tendency which ensures he will not upset himself by losing too much nor upset his opponent by winning too often.*"

Chapter 2
The Remarkable Streak of Joe Di Maggio

To gain more insight into the nature of randomness I'll begin with an informal account of what is known as the *Poisson distribution*, named after the nineteenth century mathematician Simeon-Denis Poisson. In later chapters we treat Poisson processes more generally but our primary interest here is the role of the Poisson as an approximation to the Binomial distribution.

Consider events (sometimes called "arrivals") taking place unpredictably during a span of time or over an expanse of space with the property that the number of such arrivals over disjoint times periods (or distinct regions of space) are independent of each other. When I say "arrivals" this is generally another way of talking about the occurrence of a "success" or a "hit" of some kind in which an occasional success happens sporadically at random among a bunch of "failures". Thus it is assumed that at most one event takes place at any particular instant or location and, as already mentioned, the probability of an onset does not depend on which interval of time or space we happen to be looking at. What we have, then, are the ingredients of a binomial distribution in which there are n Bernoulli trials in which the random variable $X_j = 1$ if an event takes place at the kth trial with a probability p and is otherwise 0 with probability $q = 1 - p$.

One defines a random variable X as a Poisson random variable if it assumes one of the values 0, 1, 2, ... with prob$(X = k) = \lambda^k \exp(-\lambda)/k!$ which I generally write as $P(k, \lambda)$. It is easily established that the parameter λ is the expected value E(X), namely $\Sigma\, k\, P(k, \lambda)$ summed over all k, and that $P(k, \lambda)$ summed over all k equals one, as it should (see the Tutorial for more details). The probability of exactly k arrivals, namely prob $(\Sigma X_j = k)$ over n trials can be approximated with increasing accuracy, as n gets larger, by $P(k, \lambda)$ where $\lambda = np$ is the average number of events or arrivals that occur in n trials.

So, for example, if an event is a misprint of some letter on a page with the probability of a misprint being some small p that is constant over all the letters and if n is the total number of letters on a page, then we have the ingredients for n Bernoulli trials. The probability of k misprints on a page is then approximated by the Poisson

E. Beltrami, *Digressions in Elementary Probability*, Quantitative Methods in the Humanities and Social Sciences, https://doi.org/10.1007/978-3-031-70051-4_2

distribution in which $\lambda = np$. In what follows I use the Poisson approximation in a slightly different manner that will be explained below.

A Poisson probability makes precise the notion that things are taking place in a totally haphazard manner without any discernible cause. The idea is this. Imagine a square surface (it can be any shape, actually, or simply an interval of time). You scatter a bunch of tiny pellets on it at random, meaning that each throw is independent of the previous toss and with no bias in favor of one part of the surface or another. Put in a slightly different way, each point on the surface has the same probability of being hit as any other. What do you see? One might think the pellets are spread out pretty uniformly on the square but, in fact, the distribution of points would look very irregular with clusters of them here and there, some small and others large, interspersed by some blank spots. More precisely, if one ignores the order in which events take place one can prove that their spread is randomly distributed according to a uniform distribution. This is established in Ross [13] and confirms the intuitive image of points strewn blindly with no prejudice as to where they land.

The average λ of a Poisson random variable is generally computed empirically. Let's say you dispersed N events helter-skelter in some region or, for that matter, along some time span. Now divide the spatial region or time span into a bunch M of smaller spatial or temporal partitions, each the same size and, depending on the application, ask for the probability that k events occurred in any one of these partitions, where k can be 0, or 1, or 2 or any number up to the total N. Denote by M_k the number of partitions with k events so that M_0, M_1, …, sum to the quantity M. Then the sample mean M_k/M is, by the Law of Large Numbers, approximately equal to the probability of k events per partition. This is conveyed by the Poisson probability $P(k, \lambda)$ that was defined above with $\lambda = N/M$ being the average number of events per partition. To make this more precise note that $N = \Sigma\, kM_k$ summed over all k. We also know from above that $M_k = M$ times $P(k, \lambda)$ and so, combining these two expressions, we obtain

$$N = M[P(1, \lambda) + 2P(2, \lambda) + 3P(3, \lambda) + \ldots] = M\, \lambda \exp(-\lambda) \\ \times \left[1 + \lambda + \lambda^2/2! + \lambda^3/3! + \ldots\right].$$

The sum within brackets is, as we know from calculus, an approximation to $\exp(\lambda)$ and therefore the two exponentials cancel leaving us with $N = \lambda M$ from which $\lambda = N/M$.

For example if a book has M pages and there are N misprints then $P(k, \lambda)$ is the fraction or percentage of the M pages having k misprints, with λ being the average number of misprints per page.

A striking application of the spatial Poisson approximation is provided by William Feller in his classic book on probability theory [4] in which one finds the data on where $N = 537$ flying bombs (rockets) fell on London during the Second World War. To test whether the hits occurred at random in accordance with the Poisson distribution the entire area of south London was divided into $M = 576$

enclaves of 1/4 square kilometers each and the Table below gives the number of sub-areas M_k that sustained exactly k hits:

k	0	1	2	3	4	5 and over
M_k	229	211	93	35	7	1

Using the Poisson distribution with M = 576 trials and 537 hits and letting λ be the number of hits per unit area, namely λ = N/M = 537/576 = .9323, and recalling that P(k, λ) is approximately M_k/M, it was found that

k	0	1	2	3	4	5 and over
M *times* P(k, λ)	226.	211.4	98.5	30.6	7.1	1.6

The close agreement between theory and data exhibited here suggests that a hypothesis of random hits cannot be excluded even though a resident in one of the hard hit areas might have felt it a strange coincidence that his neighborhood was singled out while other parts of London went unscathed, whereas a resident of one of these more fortunate areas might have reasoned that her vicinity was spared because that is where the enemy agents were hiding. The unexpected clustering of hits may appear suspicious even though such bursts of activity are characteristic of random processes. What a close fit to the Poisson distribution showed is that the propensity for bombs to hit one part of the city is the same as for any other part. The clustering one observes is simply a signature of chance.

Let's now ask about non-coin-tossing success runs. There are at least two examples of this kind that continue to invite controversy, namely streaks in sports and among fund managers. Since both involve people and not inanimate coins or pellets chance is often interlaced with skill and if one sees a streak of successes it is tempting to attribute that not to dumb luck but to some innate talent.

A *streak* in baseball is a succession of games in which a player has at least one hit. A *slump*, by contrast, is a cluster of games with no hit at all which, for a good hitter, may be surprising. And yet, periods of sub-par performance and other periods of extraordinary hitting may be totally compatible with simple chance and require no special explanation. The same applies to fund managers who consistently beat the market average over a succession of years and then have their own slump. A notorious example of successful hitting was Joe DiMaggio's 56 game streak in the 1941 season out of the 139 games he played. Was this an anomaly? I give a brief treatment here of a study carried out jointly with Mendelsohn [14].

Using the game by game statistics which, in DiMaggio's case, had to be painstakingly culled from newspaper box scores it is possible to count how many games in the season there were either no hits, or one hit, or two hits, up to four hits (the maximum for any game, as it turned out). Whether this data is consistent with the Poisson distribution is easily found by computing the probability P(k, λ) of k hits for k = 0, 1, 2, 3, 4 and then comparing this to the actual hits. To test whether DiMaggio's 56 game hitting streak in 1941 was an exceptional occurrence or not we made the comparison between theoretical and actual data in two ways, one with all M = 139 games accounted for and the other for the 83 games that exclude the streak. M_k denotes the number of games with k hits, N = 193 total hits in all games, N = 102 outside the streak, and $\lambda = 193/139 = 1.39$. Without the streak $\lambda = 102/83 = 1.23$. Here are the results:

All 139 games:

k	0	1	2	3	4
M_k	24	64	31	13	7
M *times* P(k, λ)	34.62	48.12	33.45	15.50	5.39

Without the streak, 83 games:

k	0	1	2	3	4
M_k	24	30	18	8	3
M *times* P(k, λ)	24.26	29.84	18.35	7.52	2.31

Looking at the first scenario we see that the Poisson assumption over-estimates the number of no hit games and under-estimates the single hit games, whereas when the streak is removed, in the second scenario, there is a very good fit to the actual data (rounding to the closest integer gives a nearly perfect fit).

The poor fit to the Poisson distribution when the streak is taken into account means, in this instance, the assumption that DiMaggio's performance can be simulated by a random coin-tossing experiment is unlikely. There are little grounds here for believing that the streak is due to pure luck. This not to say that an unusual streak cannot occur by chance alone but the probability of this happening is miniscule. We can test this by using Theorem 1 (Chap. 1).

To employ Theorem 1 (Chap. 1) we need the probability p of getting at least one hit in a game. Ignoring the streak, which was an anomaly, DiMaggio was at bat 319 times and got 102 hits for a *batting average*, usually abbreviated as BA, of .320. The number of at bats in all 139 games was 542 giving an at bat to game ratio of 541/139 = 3.9 and so the probability of no hits in any game is estimated to be $(1 - BA)^{3.9}$ assuming that each at bat is an independent event. Therefore the probability p of at least one hit per game is $1 - (1 - .320)^{3.9} = .78$. We can compare this number with his actual stats. In the 139 games he played DiMaggio got no hits in 24 of them and so his probability of getting at least one hit per game is 1 – 24/139 = .81. Close!

Under the interpretation of pure chance (luck) as being p = ½ DiMaggio's performance of p > 1/2 can be seen as a mix of luck and skill and we model his sequence of hits as 139 independent biased coin tosses with p = .78. Using Lemma 1 a streak of length at least 56 has a probability of once in 7949, a tiny chance and so the alternate hypothesis that the streak is a sort of freak is more in keeping with the Poisson model of random behavior. A hint that the Poisson approximation is appropriate, incidentally, is based on the known fact that mean and variance of a Poisson variable are both equal to λ. In our case, the non-streak value of λ = 1.23 corresponds to a variance, obtained from the data, of 1.18 which lends support to the Poisson model of randomness. Though the evidence for skill over luck is overwhelming during the streak in the other games Di Maggio played that season there is little to distinguish between the hits and misses from a sequence of heads and tails of a randomly tossed fair coin.

As a footnote to Joe's streak consider that major league baseball has about 30 teams playing each season with roughly ten players on each team having an opportunity to get a hit while at bat and there are many such seasons (baseball began in the nineteenth century). So it is not unreasonable to speculate whether among 2000 players of his caliber someone could duplicate his feat. As was seen, the probability of a streak of length 56 assuming 139 independent coin tosses (games) of a biased coin with a probability .78 of getting a head (a hit) in any game is only 1/7949. With 2000 seasons the probability we want is given by the binomial distribution: The Poisson distribution can be used to approximate the binomial from which the required computation becomes one minus the probability of getting no hits in 2000 trials. The expected value of the binomial is $np = 2000/7949 = .2516$ and this equals λ in the Poisson approximation. We therefore obtain $1 - \exp(-.2516) = .22$, a roughly 20% chance of at least one player in the group of identical players being able to achieve Di Maggio's streak. Seen this way it is a bit surprising that it has not been duplicated over the many seasons of baseball although some players have gotten notable streaks since 1941 such as Pete Rose who delivered 44 successive games with hits in 1978.

Chapter 3
Inherited, Not Acquired

Early in the twentieth century it was noticed that tiny percentages of certain strains of bacteria would continue to survive when an entire colony was placed in a toxic setting. Although it was generally understood that random mutations provide the raw material for natural selection among higher organisms it was still controversial *whether the ability of some bacteria to survive a hostile environment was due to the ability of a few to adapt to the killer virus or whether favorable random mutations that occurred all along prior to the toxic exposure is what allowed a few to survive*. In a landmark paper published in 1943 biologists Luria and Delbrück [24] successfully carried out some experiments to resolve this question in favor of spontaneous mutations. In effect they vindicated the idea that even with micro-organisms evolution works "not as a designer but as a gardener", an apt quote attributed to Jeff Bezos by Thomas Friedman (*New York Times*, May 19, 2012). The details of their investigation are given below. The key insight for interpreting their results is a very simple property of the Poisson distribution, namely that the mean and variance are the same, and this is the reason why this is of interest to us.

More specifically, Luria and Delbrück grew a large culture of bacteria (*E Coli* B), cultivated in a flask from a single cell over many generations (doubling in population at each generation), and then exposed it to a bacteriophage (virus). The virus attacks the bacteria and most are lysed. However, a few survived and gave rise to colonies that are resistant to the virus. What was not immediately obvious is whether the cells acquired their immunity by chance when the virus attacked, just as certain humans survive a massive epidemic of some deadly disease, or if the immunity was there all along as a result of prior favorable mutations. In he first instance the survivors begin to produce resistant daughter cells that form small colonies that are then counted. Because there are billions of susceptible cells each of which has the same tiny probability of mutating, independent of any other cells, this gives rise to a Binomial

E. Beltrami, *Digressions in Elementary Probability*, Quantitative Methods in the Humanities and Social Sciences, https://doi.org/10.1007/978-3-031-70051-4_3

distribution of immune cells just after they were impregnated with the harmful phage and this is the reason why it can be approximated by a Poisson distribution.

On the other hand, if mutations occur spontaneously all the time some cells become resistant before the virus is applied. The random mutations can happen in any of the previous generations and if they occur early on there will be many more resistant cells when the phage is introduced than if the mutation happened later during the incubation of the bacterial culture. In this case, if the bacterial culture in a single flask is haphazardly sampled and plated on individual plates one obtains a roughly constant fraction of resistant colonies on each plate and the Poisson fit applies whether or not one has random mutations to acquired immunity at the moment of phage contact. On the other hand if a bunch of cultures are incubated in separate flasks and then individually plated the number of surviving colonies from experiment to experiment will be much larger since a mutation that took place early during the incubation will result in a large number of immune bacteria, all of them clones of the initial mutated cell which has now divided into many generations of daughter cells (clones are survivors of a single mutant ancestor.)

Luria and Delbrück actually grew a large number of independent cultures from a single cell in separate flasks and then counted the number of mutant colonies just after being exposed to the virus. If mutations occur only on contact with the phage the distribution of survivors from flask to flask should be roughly Poisson, as we said. However, if mutants occur haphazardly in each generation as the culture is growing then the variance between flasks at the end would be much larger than the mean, and this is what they actually found. Think of each batch of cell culture as a separate baseball game in which there are 0, 1, 2, ... number of hits. Similarly, the individual flask cultures will have 0, 1, 2, ... surviving colonies of bacteria. In baseball, the mean is the number of hits per game over the season and for bacteria this quantity is the number of surviving colonies among the total number of experimental batches.

In fact, from 20 separate flasks 11 had no survivors but in the others the extant colonies ranged from 1 to 107 with a mean of 11.3 and a variance of 694, a far cry from what one expects from the Poisson distribution. The data from the original paper is shown below. To paraphrase the opening title (from a source that I can no longer recall) "Nature is an Editor and not a Composer".

In the table below we see both scenarios in action. The alternatives are between a mutation that is an adaptation to an induced challenge to survival or as it occurs spontaneously throughout the growth of the culture before being treated with the phage. The simple attribute of Poisson processes that their mean and variance are both equal was used previously in discussing baseball streaks.

Separate cultures		Single large culture	
Culture #	# of mutants	Sample #	# of mutants
1	1	1	14
2	0	2	15
3	3	3	13
4	0	4	21
5	0	5	15
6	5	6	14
7	0	7	26
8	5	8	16
9	0	9	20
10	6	10	13
11	107		
12	0		
13	0		
14	0		
15	1		
16	0		
17	0		
18	64		
19	0		
20	35		
Mean	11.3	Mean	16.7
Variance	694	Variance	15
Var/mean	60.8	Var/mean	0.9

Chapter 4
A Subtle Bias

I'll introduce the topic through some informal arguments and then go on to a formal proof. An alternate take will appear in the section on Restricted Choice using Bayes Theorem. In the subsequent discussion I follow the reasoning and even some of the wording in the striking paper by Miller and Sanjurjo [5] rather closely.

If I ask for the probability that the immediate successor of a randomly chosen head in a sequence of tosses is also a head we know, from the independence of tosses, that either a head or a tail occurs with equal frequency for a fair coin. Now I'll introduce a conundrum which I take from Miller and Sanjurjo in which we find a person tossing a balanced coin in which there is an equal chance of getting a head or tail on each flip. He generates a relatively short sequence of such tosses and then looks at those flips that follow a head. For example, if we have THTH then T follows H whereas in HHTH a head follows a head once and tail follows a head once. The flips that immediately follow a head run of length one must be a tail. One expects a considerable number of such singletons as we know from Theorem 2 (Chap. 1). For head runs of length two a head succeeds a head just once and a tail once and so on for longer (but also rarer) runs. This leads to an over-representation of tails following short runs. If T follows an H then there is no HH but if the subsequent flip is a H then the HH percentage is not always one because the second H triggers the outcome of a third flip which is equally likely to be H or T and so the probability of HH conditional on the first H will be less than 1/2. In effect we have a selection bias that forces one to consider yet another flip whenever we get HH.

To decipher what is going on let's define $P^{\wedge}(H||H)$ *as the fraction or percentage of heads that follows a head in a randomly chosen sequence of length n assuming that there is a head to choose from among the first n – 1 tosses* (such sequences are called *admissible* and they exclude a sequence of all tails or a sequence of n – 1 tails followed by a head) In effect we compute the empirical probability of HH's among all the flips that follow a head in a given sequence of tosses, given that there are heads available to follow. So, for example, in the sequence HTTH there is no HH (zero percentage) while for HHTT there is a single HH among the two heads and a single HT for an average of ½. With n flips there are 2^n possible sequences, all

E. Beltrami, *Digressions in Elementary Probability*, Quantitative Methods in the Humanities and Social Sciences, https://doi.org/10.1007/978-3-031-70051-4_4

having the same probability of occurrence since the coin is fair. For each of these one calculates the HH percentage and these are then averaged over all admissible sequences to obtain E(P^(H||H), the expectation of P^(H||H) conditional on having sequences that are not tails in the first n – 1 positions. What will be shown is that this expectation is less than 1/2 and so P^(H||H) is a biased estimator of P(H||H) = P (H) = 1/2.

The table below illustrates this for n = 3 tosses in which the first column lists the eight possible sequences of heads and tails and the second column counts how many times a head is followed by an H or a T. The third column is P^(H||H), namely the fraction of flips in column two that are actually HH. Then averaging the last column over all six admissible sequences, each counted equally, one gets 5/2 divided by 6 or 5/12 which is less than 1/2. The goal now is to establish this sampling bias in general and draw consequences in a subsequent section.

3-flip sequence	# of recorded flips	proportion of HH in column 2
TTT	0	
TTH	0	
THT	1	0
HTT	1	0
THH	1	1
HTH	1	0
HHT	2	1/2
HHH	2	1
Expected Proportion:		5/12

The procedure just described has an equivalent interpretation in terms of the probability of choosing a head that follows a head among the first n – 1 tosses of a randomly selected sequence (namely the percentage of heads that follow a head) conditional on there being a head available. If there are k available heads these are uniformly distributed among the first n – 1 positions since each position is equally likely in the sequence. If F is the event that the first n – 1 flips are not all zero (the sequence is admissible) and if A is the event that a randomly picked head among the first n – 1 tosses is followed by another head then the conditional probability P(A||F) is obtained by summing the average number of HH's over k, for k running from 1 to n – 1 and this provides the required expectation E(P^(H||H). We are using the fact that the probability of an event over equiprobable choices is a simple arithmetic average and, since this procedure computes the average HH percentage over all admissible sequences, it replicates the earlier computation of the conditional expectation of the HH percentages, namely the arithmetical average across the equally likely available sequences. The probability P(A||F) will be derived later and shown to be less than ½.

Giving equal weight to sequences of flips rather than to the flips is problematic because taking an un-weighted average of averages is an unsound statistical procedure. To give a homely example, suppose that in a community of a thousand families 900 have an average annual income of $100,000 and the other, more affluent,

100 have an average annual income of \$250, 000. An unweighted average of the incomes of both groups is then (100,000 + 250,000)/2 = \$175,000, a gross distortion. It is better to use a weighted average, that accounts for the percentage of families in each bracket. One then gets 100.000 × .9 + 250,000 × .1 = \$115,000 and this is more representative. See also Gusfield [10].

So if one focuses on the sequences and not on the individual flips we obtain a biased average. The reason is that some sequences have more heads than others, forcing heads to run together in clumps and this leads to a larger percentage of HH. This implies that sequences that have more recorded flips will tend to have a higher proportion of heads among successor flips. Because sequences that have more recorded flips are given the same weight as sequences that have fewer, any recorded flip in such a sequence will be weighted less which means that heads are weighted less. Flips following a head in sequences with fewer heads are more likely to be a tail and this is a source of the bias.

Note, however, that although our estimate for the probability of HH is biased it does not negate the irrefutable fact that H will follow a head just as often as T does. In the Table above, there are an equal number of HH and HT, four of each. If one had picked a flip at random instead of a sequence then, since the distribution of heads is uneven, the sequences would not have been chosen with equal likelihood and the HH frequency would have been a weighted average; in this case the probability of HH is indeed ½. Note that the set of values for P^(H H) in this table are either 0, ½, or 1 and the number of sequences that lead to each proportion is a skewed distribution of 3/6, 1/6, and 2/6 respectively.

I coded the computation of the probability of HH in MATLAB and it is available in the Appendix as *Millersingle.m.* The code was run 10000 times for a sequence of length 3 and p = 1/2 to obtain .417 which equals the theoretically correct value of 5/12 to three places.

The next theorem establishes a core result underpinning the bias of P^(H||H) as given in a paper by Bar-Hillel and Rinott [9]. Their proof applies to strings of any length and is valid, moreover, for biased coins having success probability p (heads) and failure (tails) q = 1 – p. The argument once again hinges on the identification of a conditional expectation of P^(H||H) with the conditional probability of the event A that a randomly chosen head from a haphazardly selected admissible sequence is followed by a head. Before getting to that I want to establish a lesser result, a sharpening of Theorem 3 (Chap. 1), that additionally supports our intuition of sampling bias.

Theorem 1 Among all sequences of n independent throws of a fair coin consider the subset in which there are m heads. Pick one of these sequences at random. The expected number of heads which follow a run of one head is (m – 1)/(n – 1) which is less than the empirical probability of m/n.

Proof Pick one of the C(n, m) sequences with exactly m heads and then select at random one of the m – 1 flips that immediately follow a head. These can be positioned from 2 to n in an equally probable manner. Thus, if F_k denotes the event that the successor flip is located at position k = 2, …, n, then

$prob(F_k) = 1/(n - 1)$. Let HH be the event that a successor flip to an H is itself an H. If this successor toss is located in positions 2 to n – 1 there are m – 2 possibilities for HH (having removed the selected trial and also the last position) and so the probability the toss is a head is (m – 2)/(n – 2). When the successor is in the last position the trial will be an H with probability 1/(n – 1).

Then $prob(HH) = \sum prob(HH \mid F_k)\, prob(F_k)$ summed over k from 2 to n – 1 and this equals $1/(n - 1) \sum [m - 2)/(n - 2)] + 1/(n - 1)$. Thus

$$prob(HH) = [(n-2)(m-2)]/[(n-1)(n-2)] + 1/(n-1) = (m-1)/(n-1).$$

If the last flip in the sequence is a tail there are now m – 1 trials available from positions 2 to n – 1 so we obtain prob (HH) = (n – 2)(m – 1)/[(n – 1)(n – 2)] = (m – 1)/(n – 1), the same quantity.

The theorem just proven is conditional on knowing how many heads there are and therefore what is needed is an unconditional probability. Instead of moving in that direction I go to a theorem whose proof looks complicated but in fact uses only the basics of Bernoulli trials and the Binomial distribution and one simply needs to pull out the strands of the argument piece by piece.

Theorem 2 The probability of the event A that a randomly chosen H among the first n - 1 tosses in a given admissible sequence is followed by another H, conditional on the probability of event F that the sequence is admissible, is given by

$$P\left(A \middle\| F\right) = (1-q)/(1-q^m) - q/m$$

for m = n – 1. Moreover, E(P^(H||H)||F) = P(A||F) < P(H||H) = p, where p is the unbiased probability that an H follows an H is any sequence and this establishes that *the estimator* ***P^(H||H)*** *of P(H||H) is biased* for any n and any p meaning that the expected value of **P^(H||H)** does not equal P(H||H).

Before giving the proof let's compute the probability P(A||F) in the case n = 3 and p = q = 1/2. The above formula gives 5/12 in agreement with what was obtained earlier.

Proof P(F) is the probability there are no T's in the first m = n – 1 positions, namely that the sequence is admissible, and evidently this equals $1 - q^m$. I now consider the event A. Suppose an admissible sequence is randomly chosen and that a head is randomly selected among the first m tosses. It is located in the last slot with probability 1/m and is followed by an H in the last position with probability p. On the other hand, if the H is located among the first m - 1 positions, with probability (m – 1)/m, then it is followed by H with probability (k – 1)/(m – 1) where k is the number of successes (heads) in m Bernoulli trials with success probability p and k ranging from 1 to m. These two events are disjoint and so the probabilities add. Moreover the union of these two events are independent of the event that a randomly chosen sequence has exactly k heads in the first m positions, the probability of which is $C(m, k)\, p^k q^{m-k}$ since there are $C(m, k)$ distinct ways of choosing the k positions. We now sum over the m disjoint possibilities for k successes, to obtain

$$P(A \| F) = P(A \cap F)/P(F) =$$
$$\Sigma C(m,k) p^k q^{m-k} [p/m + (k-1)(m-1)/m(m-1)]/(1-q^m)$$

The sum over k of $C(m, k)\, p^k q^{m-k}$ from 0 to m is one and therefore the sum from 1 to k is $1 - q^m$. Also the expected value of the binomial distribution is mp and so, using these basic facts, it is a straightforward computation to arrive at

$$P(A \| F) = (1-q)/(1-q^m) - q/m = [m(1-q) - q(1-q^m)]/m(1-q^m)$$

But $(1 - q^m)/(1 - q) = 1 + q + \ldots + q^{m-1}$ and this becomes

$$\left[m(1-q) - q(1-q)\left(1 + q + \ldots + q^{m-1}\right)\right]/m(1-q^m) \tag{4.1}$$

Subtracting (4.1) from $p = 1 - q$ we obtain

$$\left[m(1-q)(1-q^m) - m(1-q) + q(1-q)\left(1 + q + \ldots + q^{m-1}\right)\right]/m(1-q^m) \text{ or}$$
$$(1-q)\left[m(1-q^m) - m + q\left(1 + q + \ldots + q^{m-1}\right)\right]/m(1-q^m) =$$
$$q(1-q)\left[\left(1 + q + \ldots + q^{m-1}\right) - mq^{m-1}\right]/m(1-q^m) \tag{4.2}$$

But the sum of terms within the parenthesis is greater than mq^{m-1} and so (4.2) is positive. This suffices to establish the bias.

I cannot resist mentioning here a well known classical *estimation bias* : let X_k be a set of n *identically distributed and independent* random variables from a probability distribution having mean m and variance σ^2. The quantity $m^\wedge = (\Sigma X_k)/n$ is an *unbiased estimate* of m, namely $E(m^\wedge) = m$. One sees this because $E(m^\wedge) = 1/n \sum E(X_k) = nm/n = m$.

Since the variables X_k are independent, $\mathrm{Var}(\sum X_k)/n = (1/n)^2 \sum \mathrm{Var}(X_k) = \sigma^2/n$ and it is reasonable to compute an estimate of variance by $S^\wedge = 1/n \sum (X_k - m^\wedge)^2$ summed over k. After a short computation one obtains $E(S^\wedge) = (1 - 1/n)\sigma^2$ which, unlike $m^\wedge$, is a *biased estimate* since $E(S^\wedge) \neq \sigma^2/n$. To correct this one needs to divide the previous sum by $n - 1$ instead of n.

Chapter 5
Hot Hands in Basketball

The selection bias for HH impinges on the topic of streaks in sports and has led to a much debated controversy in basketball concerning the phenomenon of "hot hands". An oft-quoted paper by Gilovich, Vallone, and Tversky [8] debunks the idea that basketball players are more likely to score a second hit after a successful shot or, to put it another way, the first and second shots are independent Bernoulli trials. In his book *Thinking, Fast and Slow* [36], Nobel economist Kahneman comments (well before the Miller-Sanjurjo paper appeared) that "the hot hand is entirely in the eye of the beholders, who are too quick to perceive order and causality in randomness. The hot hand is a massive and widespread cognitive illusion." Nevertheless, most spectators and players believe that streak shooting is a true phenomenon. The results of the Gilovich et al. study remained controversial until Miller and Sanjurjo shed new light on the question by showing that Gilovich et al. underestimated the phenomenon of "hot hands" or "being in the zone".

Note that a streak in basketball is about getting a hit immediately following a given sequence of successes and the percentage of successes are then averaged over a number of such sequences within a game or over several games whereas a streak in baseball is a single unusual run of hits within the successive games of a single season. In each instance it is possible to speculate that what we observe is due solely to chance, a sequence of Bernoulli trials.

Before Miller-Sanjurjo several researchers had tried to salvage the hot hands idea by saying, for example, that success did increase the chance of subsequent success though generally over a larger time frame than the next shot. To quote from Reynolds in the NY Times [17]: "basketball players experienced statistically significant and recognizable hot periods over an entire game or two, during which they would hit more free throws than chance would suggest. But they would not necessarily hit one free throw immediately after the last. . .in fact they are more likely than chance would suggest to miss the next throw". This is the case with DiMaggio's streak which occurred over subsequent games but not over successive at bats within a game. In fact during the streak he got a single hit in 34 out of the 56 games he played. So a single hit was followed by a miss with probability of about $34/56 = .60$

E. Beltrami, *Digressions in Elementary Probability*, Quantitative Methods in the Humanities and Social Sciences, https://doi.org/10.1007/978-3-031-70051-4_5

which means that the likelihood of a hit following a single hit was only 40 %. Thus the streak was seen only at the level of successive games and not within a single game.

An analogous explanation, given by Wardrop [16], invokes the curious statistical fluke known as Simpson's Paradox in which probabilities of two events are reversed when the events are collapsed into one. The example he offers is that of two players from the Boston Celtics, Rick Robey and Larry Bird, during the 1980–1981 and 1981–1982 seasons. Each player had a slightly better chance of getting a hit after a miss compared to a hit after a hit but when one combined their individual scores into one, disregarding the identity of the players, the computed probability of getting a hit after a hit became substantially better than getting a hit after a miss. Wardrop interprets this as a manifestation of a cognitive illusion that after observing multiple players over many seasons of games the appearance of hot hands is the mental equivalent of Simpson's paradox that results in "seeing patterns in the data that do not exist." In the hustle and bustle of actual games the spectator perceives streaks while the statisticians record independent Bernoulli trials. I will have more to say about Simpson's Paradox in a later vignette where I will analyze the Boston Celtics data in considerably more detail.

A similar comment is made by Dorsey-Palmateer and Smith [20] who note that there are many confounding variables in evaluating field goals during actual games and, moreover, the time between successive shots for a player can be a matter of seconds, minutes, in different halves of a game, or even between games, so the notion of being "in the zone" becomes somewhat ambiguous. A more controlled environment for the evaluation of hot hands would be the NBA three-point contest.

A controlled experiment was in fact carried out by Gillovich et al. [8] who looked at the difference D between a player getting a hit after three consecutive hits minus getting a hit after three consecutive misses during 100 consecutive throws. That is, D records the difference $P \wedge (H||3H) - P \wedge (H||3T)$ for each player. The players were 26 members of Cornell basketball teams who shot from positions deemed likely to give a 50% success rate. The D percentage for each individual were mostly small plus and minus values and when averaged over all players they found a statistically insignificant $D = .03$. Because they unwittingly assumed this to be an unbiased randomized procedure it was concluded that the effective indifference to success after successive hits or after successive failures indicates that the players did not exhibit hot hands; The sequence of throws were deemed to be independent trials like tossing a coin. Thus hot hands was ruled out as an illusion of players and fans who are accustomed to seeing patterns in randomness and interpret an occasional success run as a streak bowing to the predisposition, discussed earlier, of thinking of chance as more frequent alterations between heads and tails than is actually the case.

Now independence means $P(H||H) = P(H||T)$ while hot hands implies that $P(H||H) > P(H||T)$. But, because of Theorem 2 (Chap. 4), $E(P \wedge (H||H) < 1/2$ which suggests that more often than not, under independence, one will observe $P \wedge (H||H) < P \wedge (H||T)$ *. Similarly, observing small plus and minus values of $P \wedge (H||3H) - P \wedge (H||3T)$, as did Gillovich et al, rather than mostly negative values, as expected under independence, suggests that independence may be violated, In fact

Miller-Sanjurjo established that if the data is truly from a sequence of 100 (independent) Bernoulli trials with a success probability of ½ then the biased estimation procedure used by Gilovich et al. should have given a probability of .46 for P ^ (H||3H) and, for reasons of symmetry, a corresponding probability of .54 for P ^ (H||3T) a significant difference of D = – .08. Since Gillovich et al. found D = .03 the actual point spread is, on average, .11 and, as stated by Miller and Sanjurjo, this is large enough to not only negate the 'hot hand fallacy' but in fact to support a contrary belief in the validity of a hot hand.

Now consider the *Gambler's Fallacy*, a belief that a long succession of heads in coin tossing will soon be reversed because the long run frequency of heads/tails tends to equalize (Law of Large Numbers). As discussed earlier this long term equalization is incorrectly viewed to hold even in relatively short runs; the feeling is that as a short run of heads increases so does the probability that the next toss is a tail. Of course we know that in independent tosses the probability of either a head or tail in a given toss of a fair coin remains ½. So, what seems to be going on is that gamblers observe many short sequences of complete games over time applying an equal weight to each of these sequences or games obtaining what is in effect an unweighted average of HH percentages. This under-estimates the true probability that an H follows an H and, as Miller and Sanjurjo comment, "thereby confirming a belief in a reversal of fortune, consistent with the gambler's fallacy."

I close with another quote from Miller-Sanjurjo: "Gillovich, Vallone, and Twersky see a binomial process of independent throws when there is in fact streak momentum while a gambler makes the mirror fallacy of seeing a reversal in a similar binomial process when in fact there is no momentum".

* Note that a head must be followed by an H or a T so that

$$P^\wedge\left(H\middle\|H\right) = 1 - P^\wedge\left(T\middle\|H\right).$$

Moreover, in Bernoulli trials with success probability of 1/2 the likelihood of H is the same as T and it is evident, using the definition of conditional probability, that P ^ (T||H) = P ^ (H||T). Then, since E(P ^ (H||H) < 1/2, it follows that on average

$$P^\wedge\left(H\middle\|T\right) = 1 - P^\wedge\left(H\middle\|H\right) > P^\wedge\left(H\middle\|H\right).$$

Chapter 6
Restricted Choice

The simplest and perhaps most persuasive argument for the selection bias was provided by Miller and Sanjurjo [6] employing Bayes Theorem and I can do no better than to follow their reasoning and even some of their wording.

Baye's result is discussed in the Tutorial at the end of the book as is the concept of odds. These ideas will come up often in this and the remaining sections.

To begin let's state Baye's theorem in odds form when considering a choice between two alternatives whose outcomes are unknown, say A and ~A. If B is the event that some observation has taken place the likelihood of B is the probability of B given that A is true, P(B||A). Baye's theorem is that P(A||B) = P(A||B)P(A)/P(B). A similar relation holds with A replaced by ~A. Dividing one by the other and defining the Bayes Factor, BF, as the ratio P(B||A)/P(B|| $\sim$ A) we then obtain Baye's Theorem in the form Odds(A || B) = BF x Odds(A).

I now consider the case of three flips of a fair coin in which we assume that there is a head among the first two tosses. A person selects at random one of the flips that immediately follows an H. The first two tosses are either HT or TH or HH, all having the same probability. In the first two instances the person picks the second or third flip and for HH picks the second or third at random uniformly. If it happens that the second flip is the one picked it means that we are faced with either HT or HH. Before flipping the odds of in favor of HH relative to HT is 1: 1 but when HT occurs the selection of flip 2 is twice as likely than when HH occurs since in the latter case one could also have chosen flip 3 whereas for HT we are restricted to flip 2. That is, BF = P(flip2||HT)/P(flip2||HH) = 1/1/2 = 2 and by Bayes Theorem P(HT||flip 2)/P (HH||flip 2) = 2 × P(HT)/P(HH) = 2. With odds of 2: 1 the probability of HT is 2/3.

On the other hand, if the third toss is chosen as the successor flip it means that TH or HH are equally likely to appear in the first two positions and in this case the odds don't change. Altogether, then, since we have two disjoint events, P(T||H) = 1/2 × 2/3 + 1/2 × 1/2 = 7/12 and so P(H||H) = 5/12, exactly as obtained earlier.

E. Beltrami, *Digressions in Elementary Probability*, Quantitative Methods in the Humanities and Social Sciences, https://doi.org/10.1007/978-3-031-70051-4_6

Now consider 100 flips of a fair coin in which we select a flip among those tosses that follow three consecutive heads. An argument similar to the one given in the previous paragraph continues to be valid. If flip 42 is selected, for example, then one is again led to think the there is an equal chance this flip is H or T. This is not the case because in the event of HHHH, flips 39-42, one could also have chosen flip 43 as a successor but with HHHT we are constrained, namely restricted, to pick flip 42 and so in this particular head run the likelihood of choosing 42 when HHHT occurs is more likely than when HHHH takes place and it is therefore more likely for 42 to have been picked initially. This applies to any of the head runs among the 100 tosses (except for the last toss). See also [15].

Remarkably, the surprising sampling bias Miller and Sanjurjo uncovered is unexpectedly related to the notorious Monty Hall Problem (see Rosenhouse [19]) and the idea of restricted choices informs this puzzle as well as shown by them in their paper [6]. Based on a television game show called "Let's Make a Deal" the host Monty shows a guest three doors having a car behind one of them and goats behind the other two. Monty of course is the only one who knows where the car is located. Our guest is offered the opportunity to open one of the doors and be rewarded by either a car or a goat. The guileless contestant chooses a door but before it can be opened Monty will open a different door to reveal a goat. Then Monty says that the contestant has the option to stick with the initial choice or to switch to one of the two remaining doors. What happens next?

Writing G for goat and C for car there were three equally-likely scenarios before Monty acted: GCG, GGC, and CGG (left to right). Monty will only open a door hiding a goat and, moreover, if two goat doors are available to him he chooses between them uniformly at random. The guest initially chooses door three, let us say. This shrinks Monty's options to two possibilities: either the car is behind door three in which case Monty is equally likely to open one of the two remaining doors or else Monty is now restricted to open the one door having a goat. Let's say he opens door one to reveal a goat. This shrinks the initial three options to just two: GCG or GGC. As far as the contestant is concerned these two options have the same chance of occurring and so the prior odds of the event A = GCG being true is one. So, given event A is true the probability of Monty opening door one is certain since he has no other choice; letting B denote the event that door one is opened this implies that Prob $(B|A) = 1$ whereas Prob $(B|\sim A) = ½$ because under this scenario ($\sim A$ = GGC) he has two options and, picking at random, the probability he choses door one is ½ . It follows that the Baye's Factor is 1 divided by ½ or 2. It is now immediate that the posterior odds of A is 2: 1. This corresponds to a probability of 2/3. It is evidently in the contestant's best interest to switch since the probability of getting a car has now jumped from 1/3 to 2/3. Put differently, the odds of GCG is two in favor to one against. A surprising result when one first encounters it! The Monty Hall problem achieved considerable notoriety when the New York Times initially reported on it as front page news [41]. Rosenhouse offers another convincing argument in his book by identifying each goat separately as G1 and G2 since this leads to six equally probable arrangements go possibilities behind each door:

Door 1	Door 2	Door 3
G1	G2	C
G2	G1	C
G1	C	G2
G2	C	G1
C	G1	G2
C	G2	G1

If the contestant chooses door 3, let us say, then since each arrangement has probability 1/3 it is fairly apparent that she wins a car by switching in 4 out of six cases with a probability of 2/3. The same is true regardless of which door was initially chosen.

An alternate, more visual approach, to understanding the Monty Hall problem is through a tree diagram and can be found in Rosenhouse's book [19] and even in a delightful novel by Mark Haddon called The Curious Incident of the Dog in the Night [74].

My interest in mentioning this confounding puzzle here, besides being another example of restricted choice, is its connection with a cognitive illusion well known to psychologists and behavioral economists as discussed in two NY Times articles by Tierney [50] and [51] as well as in the lucid and entertaining account by Rosenhouse [19]. The idea is that once a contestant has chosen a door he or she is loathe to switch on the grounds that an individual who switches and gets a goat experiences regret and is more likely to harbor the unsettling thought, to quote Rosenhouse, "I brought this on myself" or "this need not have happened" compared to someone who decided to keep their original choice. It's a worst feeling to switch and lose than to stick to the initial choice and lose since in the latter case one can shrug off the loss by thinking "it wasn't meant to be". One uses self delusion to avoid the mental discomfort, the *cognitive dissonance*, of thinking we made the wrong choice. However, in spite of such misgivings the mathematical consequence of Baye's theorem is independent of these cognitive discomforts and the conclusion that it is best to switch remains true regardless. We will see a very similar situation in the next paragraph.

A striking example of how we rationalize choices is a simple example of providing a monkey, say, with three identical candies that come in three colors red, blue, and green, R, B, G. Assuming that the monkey is indifferent to the color of the candy randomly pick two of them, say R and B, and let the monkey choose one. If B is rejected then, in a subsequent choice between B and G, the monkey would prefer G most of the time in order to avoid cognitive dissonance according to some researchers. The blue candy is devalued, so the theory goes, which is reminiscent of the Monty Hall quandary.

An analysis like that of restricted choice shows that having randomly and uniformly decided between R and B there are now three possible rankings that preserves the preference for R over B: RBG, RGB, and GRB. A prior indifference between R and B now gives a posterior probability of G being preferred to B of 2/3 since there is only one possibility for picking blue over green, the restricted choice of RBG.

However, this resolution of the monkey's choice does not depend on an assumption of cognitive dissonance any more than it impacted the solution of the Monty Hall problem.

The unexpected outcomes introduced by restricted choice is illustrative of troublesome illusions induced by the often counter-intuitive nature of chance events. A wide array of these surprising results are encountered in this book. Some others are introduced by Kahneman in his influential book [36] but many of them were already recognized early in the nineteenth century by Pierre Simon Laplace who contributed to many parts of mathematics including probability theory. His insights in probability are recounted in a paper by. Gelman and Miller [33].

Chapter 7
Another Conundrum: What Does the Crowd Size at Wimbledon's Tennis Match Tell Us About Baseball Batting Averages?

I want to discuss a curious finding that appears to defy intuition. I begin by reviewing a classical estimation procedure in which we have a random variable having a normal density function with unknown average or mean θ and a unit variance. What needs to be estimated is the unknown mean θ.

A set of n independent samples of the normal random variable are denoted by X_1, X_2, . . ., X_n, The basic properties of a normal density function are reviewed in the Tutorial and it is recommended that one take a look at that material before proceeding.

Let $X = (X_1, X_2, \ldots, X_n)$. It is known, as will be shown later in the Addendum to this section, that the *best unbiased estimate of the mean* θ in terms of minimizing the *mean squared error* is actually the sample mean $X \wedge = 1/n \sum X_j$ summed over j. More precisely, suppose that a function d(X) of the normal variables X minimizes the expected squared difference between d(X) and θ, namely it minimizes $E(d(X) - \theta)^2$ and, moreover, it is unbiased in that $E(d(X)) = \theta$. Since it is unbiased $E(d(X) - \theta)^2$ is the variance of d(X). It turns out, as I'll show in an Addendum, that this best unbiased estimate d(X) is in fact X^. It is clear that X^ is unbiased since $E(X_j) = \theta$ for all j. Then $E\,[(X \wedge - \theta)^2]$ is the variance of X^.

Now consider a problem with k unknown means θ_j, j= 1, 2, . . ., k, from normal densities in which we have formed, as discussed above, k sample means X_j (now denoted without the use of ^ since the sample means are approximately normally distributed if the sample size is large enough as discussed in the Tutorial). These are the best estimates of the *individual* means θ_j meaning that $E\,[(d(X_j) - \theta_j)^2]$ becomes as small as possible. It would appear that the sample means would again be optimal *collectively* for $\sum E\,(d(X_j) - \theta_j)^2$ summed over k, *a quantity generally referred to as the risk*. But this is not the case when $k \geq 3$ as I'll explain in what follows, giving me an opportunity to introduce a surprising and counterintuitive statistical estimation procedure.

An example, carried out in detail by Efron and Morris in [44], is that of 18 major league baseball players in the 1970 season in which X_j are the individual batting averages over the first 45 at bats of the season. The idea is to take the batting

E. Beltrami, *Digressions in Elementary Probability*, Quantitative Methods in the Humanities and Social Sciences, https://doi.org/10.1007/978-3-031-70051-4_7

averages of the players at the beginning of the season and use these to estimate the true but unknown empirical probability θ_j of each player getting a hit (namely the number of hits divided by the number of times the player is *at bat*) over the length of an entire season; each time a player is at bat is a Bernoulli trial with success probability of getting a hit equal to θ_j. The best and worst performers early on will regress towards seasonal averages that can be quite different from their initial values.

However, a curious fact emerges in that for at least three players there is even a better estimator Z_j of θ_j for which the overall risk is less. This surprising result is known as *Stein's Paradox* following Stein's paper [46]. What Stein proposed is to estimate the unknown θ_j by $Z_j = \underline{X} + (1 - b)(X_j - \underline{X})$ where $\underline{X}$ is the mean of the means, namely $(\Sigma X_j)/n$ summed over the k players, and b is a suitable positive constant less than one. In effect each X_j is shrunk towards a common mean $\underline{X}$ over all players. For baseball $\underline{X}$ is the average of the seasonal averages of all the players. Remarkably, $\Sigma E(Z_i - \theta_i)^2 < \Sigma E(X_i - \theta_i)^2$ when there are three or more players.

I list below six of the 18 players with Clemente, a top hitter, first and Alvis, the low scorer (based on the first 45 at bats) being last. The batting averages X are the mean square estimates (over 45 trials) of the seasonal averages θ while Z are the Stein estimates:

Name	X	θ	Z
Clemente	.400	.346	.290
Johnstone	.333	.222	.277
Kessinger	.289	.263	.270
Petrocelli	.222	.264	.254
Swoboda	.244	.230	.261
Alvis	.156	.200	.239

We see that Stein's approach underestimates Clemente's true ability and overstates that of Alvis. Nonetheless, it can be shown [44] that when averaged over all 18 players the Stein estimates gives a substantially lower risk than the one generated by mean square error. This is achieved by penalizing exceptional players; the minimum takes place with no concern for individual cases. All players have scores that are shrunk towards the common mean $\underline{X}$ which, in this case, turns out to be .265. There is a close relationship between this shrinkage and the notion of regression to the mean which is explored by Stigler in [34] and there is also a strong connection with Bayesian estimation. I do not pursue that here.

But what makes this estimation procedure truly bizarre is that the θ's can come from totally different populations. In [44] for example the baseball averages are mingled with a sample mean of imported vehicles among 45 randomly chosen automobiles in Chicago or, as mentioned in the Wikipedia article [46], one can add to the *Stein estimator* three unrelated and independent estimates of the US wheat yield for 1993, the number of spectators at the Wimbledon tennis tournament of 2001, and the weight of a randomly chosen candy bar. Then the Stein estimator results in a better overall estimate of the unknown means that is possible using a mean squared error risk. But what can the spectator size at Wimbledon in 2001 possibly have to do with baseball batting averages in 1970? And, for that matter, what does the batting average of the highest scoring player have to do with the performance of the least scoring player? It is hard to fathom any link. As one exasperated author wrote "this is just plain nuts' [45].

In a short note by Samworth [75] regarding the Stein estimator we read that "if we are interested in the proportion of the US electorate who will vote for Obama, the proportion of babies born in China that are girls and the proportion of Britons with light-colored eyes then our James-Stein estimator of democratic voters depends on our hospital and eye color data!"

But, as the Wikipedia article goes on to say "This is of course absurd; we have not obtained a better estimate for wheat yields itself'. What has been obtained is a reduced risk for the random variables *when averaged together.* This total risk reduction can lead to a worst estimate for any individual variable compensated by a better estimate for another component.

In the following I use a modification of Stein's estimator given by $Z_j = (1 - 1/S^2)$ X_j where, as before, the X_j are n individual normal sample means of unknown population means θ_j and $S^2 = \sum X^2_j$ summed from 1 to n. For simplicity the mean of means $\underline{X}$ is taken to be zero. The $1/S^2$ term is called the *shrinkage* factor.

The proof of Theorem 1 (and the Addenda that follows it) is entirely optional and is included for the curious reader who is conversant with multivariate normal random variables. Note that I use the single integration symbol $\int$ to indicate the n - fold integral $\int \int \ldots \int$.

Theorem 1 $E(\sum (Z_j - \theta_j)^2 < E(\sum (X_j - \theta_j)^2$ for $k \geq 3$

Proof Begin with the identity $(X_j - Z_j)^2 = (\theta_j - Z_j + X_j - \theta_j)^2 = (\theta_j - Z_j)^2 + (X_j - \theta_j)^2 - 2(Z_j - \theta_j)(X_j - \theta_j)$. Transposing one gets

$$(Z_j - \theta_j)^2 = (X_j - Z_j)^2 - (X_j - \theta_j)^2 + 2(Z_j - \theta_j)(X_j - \theta_j). \qquad (7.1)$$

Taking expectations of (7.1) and summing we obtain, recalling that $Cov(Z_j, X_j) = E[(Z_j - \theta_j)(X_j - \theta_j)]$, one finds

$$\sum E(Z_j - \theta_j)^2 = \sum E(X_j - \theta_j)^2 - \sum (X_j - \theta_j)^2 + 2\sum Cov(Z_j, X_j). \qquad (7.2)$$

Since it was stipulated that the variance of $X_j = 1$ the second term on the right of (7.2) equals n. The unit variance of X_j together with their independence permits us to define their joint density function as $f(x_1, x_2, \ldots, x_n)) = 1/(2\pi)^{n/2} \exp(-1/2 \sum (X_j - \theta_j)^2$ in terms of the unknown means θ_j.

$$\text{From this it follows that } \partial f / \partial X_j = -(X_j - \theta_j)\, f(X). \qquad (7.3)$$

Using the definition of covariance (reprised above) (7.3) gives

$$2\mathrm{Cov}(Z_j, X_j) = 2\sum \int (Z_j - \theta_j)(X_j - \theta_j) f(X) =$$

$$-2\sum \int (Z_j - \theta_j)\ \partial f/\partial Xj. \tag{7.4}$$

Now integrate (7.4) by parts to obtain $2\sum \int \partial Z_j/\partial X_j\, f(X) = 2\sum E(\partial Z_j/\partial X_j)$ from which we find

$$\sum E(Z_j - \theta_j)^2 = \sum E(X_j - Z_j)^2 - n + 2\sum E(\partial Z_j/\partial X_j).$$

Recall that

$$Z_j = (1 - 1/S^2)X_j \tag{7.5}$$

Using (7.5) it follows that $\sum E(X_j - Z_j)^2 = E(\sum(X^2_j/S^4)) = E(1/S^2)$. Moreover, again using (7.5),
$2\partial Z_j/\partial X_j = 2 - 2\partial X_j/S^2)/\partial X_j = 2 - 2(E(1/S^2 - 2X^2_j/S^2)$ and therefore

$$2\sum E(\partial Z_j/\partial X_j) = 2n - 2(E(n/S^2) - E(2/S^2)) = 2n - 2(n-2)E(1/S^2).$$

Altogether, then,
$\sum E(Z_j - \theta_j)^2 = n - 2(n - 2)E(1/S^2)$. The last term on the right is positive for $n \geq 3$ which proves the theorem.

Addendum

Let $X_1, X_2, \ldots, X_n$ be independent samples from a normal density having unit variance and mean θ. Define an *unbiased estimator* of the unknown θ as some function $d(X) \equiv d(X_1, X_2, \ldots, X_n)$. The problem is to choose this estimator with $E(d(X)) = \theta$ so that it minimizes the risk $E((d(X) - \theta)^2) = \mathrm{Var}(d(X))$.

The joint density of the X variables is

$$f(x) \equiv f(x_1, \ldots, x_n) = 1/(2\pi)^{n/2} \exp\left(-1/2\sum (x_j - \theta)^2\right)$$

I want to establish that $\mathrm{Var}(\sum X_j/n) \leq \mathrm{Var}(t(X))$. By definition $E(\sum X_j/n) = \int (\sum x_j/n) f(x)dx = \theta$ and we also have that

$$E(t(X)) = \int t(x)f(x)dx = \theta \tag{7.6}$$

Now differentiate (7.6) with respect to θ to obtain

$$\int t(x)\left(\sum(x_j - \theta\right)f(x)dx = 1 \tag{7.7}$$

Also, since

$$E\left(\sum(X_j - \theta\right) = 0,\ \theta \int \sum(x_j - \theta)f(x)dx = 0. \tag{7.8}$$

Now subtract (7.8) from (7.7) to find
$\int(t(x) - \theta)[\sum(x_j - \theta)]f(x)dx = 1$ or, letting $T = t(x) - \theta$ and $S = \sum(x_j - \theta)$,

$$\int T(x)S(x)f(x)dx = 1 \tag{7.9}$$

Let's apply the *Cauchy-Schwarz inequality* to (7.9) (I'll prove this inequality in a second Addendum below):

$$1 = \left\| \int T(x)S(x)f(x)dx \right\| = \left\| \int T(x)\sqrt{f(x)}S(x)\sqrt{f(x)}dx \right\| \le$$

$$\left[\left(\int T^2(x)f(x)dx\right)\right)\left(\int S^2(x)f(x)dx\right)]^{1/2} \tag{7.10}$$

Now $\int S^2(x)f(x)dx = E(S^2)$. But $\mathrm{Var}\, S = E(S^2) - [E(S)]^2$ equals $E(S^2)$ since $E(S) = 0$. That is, $E(S^2) = \sum \mathrm{Var}(X_j - \theta) = n$. It therefore follows from (7.10), after first noting that

$$\mathrm{Var}\left(\sum X_j/n\right) = n/n^2 = 1/n,$$

$\int T^2(x)f(x)dx \ge 1/n = \mathrm{Var} \sum X_j/n)$. But $\int T^2(x)f(x)dx = E(t(X) - \theta)^2 = \mathrm{Var}(t(X))$ and so $\mathrm{Var}(t(X)) \ge \mathrm{Var}\ (\sum X_j/n\ )$, as required.

Addendum: Cauchy-Schwarz Inequality

The previous addendum used an inequality that I'll establish here. Again, purely optional (it uses calculus).

Assume functions f and g and their squares are integrable from 0 to ∞. For any scalar λ consider the quadratic in λ given by the inner product of $(f + \lambda g \int fg\, dx)$ with itself. This is non-negative and therefore all roots of the quadratic are either complex numbers or there is a multiple real root. This implies that the discriminant of the quadratic cannot be positive. Expanding out the inner product we see that this discriminant condition is equivalent to the desired inequality, namely $(\int fg\, dx)^2 \leq \int f^2 dx \int g^2 dx$.

Chapter 8
Cancer Clusters

It is not inconsistent with randomness to have regions in the United States in which there is an unusually high incidence of some specific cancer. These *cancer clusters* occur in various parts of the nation and lead to a belief among some residents of these communities that there must be an unnatural source for the higher than usual rate of malignancies, such as toxic wastes secretly dumped into the water supply by local industries or government agencies who connive in a conspiracy of silence. It seems like just too much of a coincidence. Public health agencies are often cajoled into investigating what the residents of such a targeted township regards with suspicion, disregarding the fact that there are many such locales throughout the nation some of which may not even be aware that their cancer rate is above average. Once again it is a case of ignoring that the probability of such a cluster, somewhere, not just in your backyard, may not be all that small. As the Poisson approximation shows, a cluster is likely to happen somewhere by chance and will affect someone other than you and your community. When it happens to you it induces skepticism and distrust (see Gawande [18]). This is a common problem of misjudging coincidences, a topic that is taken up in more detail in a later vignette.

I simulated this phenomenon by randomly dispersing 80 cases of a rare cancer in a region of 64 square miles, the size of a moderate town. This region is then further subdivided into 64 neighborhoods of one square mile each some of which contain no incidents (within a year, say) and other contingent neighborhoods having as many as four cases. This is shown in the figure below which was obtained by a short MATLAB code:

```
x=rand (1,80)*64;
y=rand (1,80)*64;
scatter (x, y)
grid on
```


E. Beltrami, *Digressions in Elementary Probability*, Quantitative Methods in the Humanities and Social Sciences, https://doi.org/10.1007/978-3-031-70051-4_8

These 80 points follow a Poisson distribution and it is plainly visible how the points clump, much as in the previously discussed rocket hits on London. In fact by counting how many neighborhoods have either zero or one or two or three and more cancer cases we can then compare this to what the Poisson distribution implies. In this example lambda = 80/64 = 1.25:

Cases:	0	1	2	>2
Data:	17	21	16	8
Poisson:	18.3	22.9	14.3	6.0

The fit is pretty good.

Instead of measuring the number of cancer cases per unit area it may be more informative to compute cases per unit population (per-capita cases). If some areas are sparsely populated and others more densely inhabited the spread within the sparse areas is considerable due to the high variability caused by the small sample size. These less dense regions exhibit morbidity rates that range from zero to numbers much higher than the population average taken over all the neighborhoods together. This is because the per-capita rates are sample means from a normal distribution that have a spread that depends on the square root of the sample size, something that will be explained more fully in the next section. As a result the more sparse areas can sustain both the lowest and highest morbidity rates in contrast to the more thickly populated neighborhoods whose rates are much closer to the national average. One would be tempted to (as, for example, by Wainer [21] with regard to the incidence of kidney cancer, county by county) *attribute low incidence to a healthy life style and high incidence to poor medical care but such explanations could be premature and quite possibly spurious, more of a statistical artifact than a causal link*. What Wainer shows is that when county population is small there is a wide variation in cancer rates from 0 to 20 per 100,000 people; when county populations are large there is little variation with about 5 cases per 100,000 for each of them. Moreover, the number of cases does not grow proportionately to population but, rather, to its square root so that, for example, the increase in incidence in a county of 100,000 people varies by about 100 which is only 10 times more than a county with just 100 people. I will comment further on this phenomenon in the next section (Fig. 8.1).

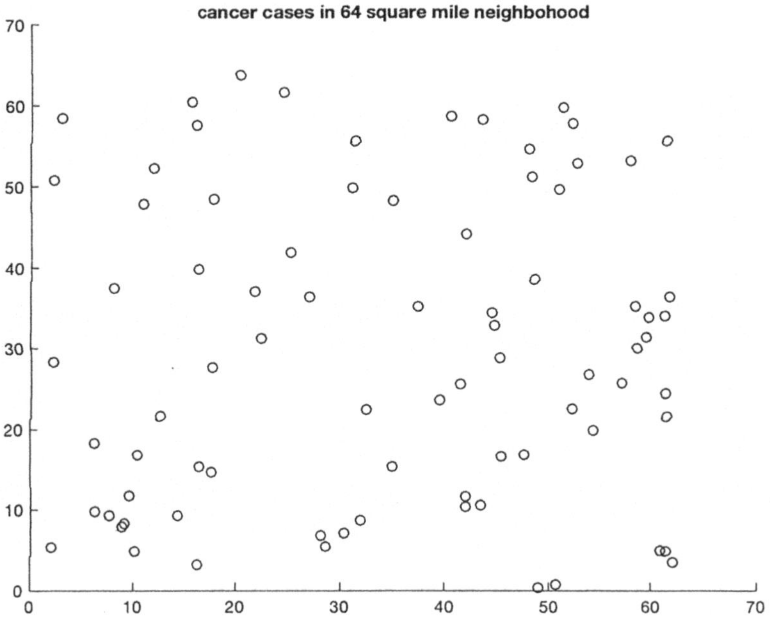

Fig. 8.1 Cancer cases

Chapter 9
Medical Misreadings

In the previous section we saw that a random sample of per-capita cancer rates by county can exhibit considerable variation depending on sample size. *The smaller the sample the larger the variance and so rural counties with less dense populations tend to have both the largest and the smallest per-capita rates whereas larger counties have a more narrow spread around the nationwide average rate. This leads one to conjure narratives to explain what is going on.*

It is easy to succumb to another statistical misuse of medical data, the *base rate fallacy*. A good example is the introduction in 2017 of a drug Evolocumab, brand name Repatha, that significantly reduced HDL cholesterol levels in patients having a high risk of a heart attack or stroke. The New York Times reported on a study [47] that appeared in the New England Journal of Medicine in 2017 with the title Evolocumab and Clinical Outcomes in Patients with Cardiovascular Disease. This was followed a few days later by another NY Times piece damping some of the enthusiasm [48].

A group of 27,600 patients already taking a statin to reduced to reduce their risk was divided into two cohorts, one assigned the drug and the other a placebo, and so 13,800 took the drug intravenously. After 2.2 years this group was found to be 20% less likely to die from heart disease or stroke. In fact 816 individuals take Repatha had a bad outcome compared with 1013 people who took the placebo. The reduction in risk is therefore (roughly) 20% of 1013. That means that the base rate for a serious cardiovascular event among the patients in the study, namely the disease prevalence within the group, is $1013/13{,}800 = .073$ or 7.3%. Among those taking the drug, however, a fraction $816/13{,}88 = .060$ were spared serious event and so the absolute risk reduction was $.073 - .060 = .013$. So, *from the viewpoint of any specific individual in the study out of every 1000 individuals 13 had a risk reduction meaning that about 77 needed to be treated in order to benefit a single person.* Not very impressive given the considerable expense and inconvenience of administering the drug over an extended period of time. The original report promoted the relative risk reduction of 20% because over a large body of susceptible individuals this appears to be an impressive accomplishment.

E. Beltrami, *Digressions in Elementary Probability*, Quantitative Methods in the Humanities and Social Sciences, https://doi.org/10.1007/978-3-031-70051-4_9

It is true that the absolute risk reduction increases to 2% over 3 years but it still means that about 50 people need to be treated with Repatha for one to benefit. If the base rate had been larger (a higher percentage of disease prevalence) then the absolute reduction in risk would have been more impressive.

For other illustrations of the base rate fallacy see the survey in [72].

Data from hospitals in England during 2005/6 were analyzed in a paper [27] regarding hospital mortality rates following certain surgical procedures. Each hospital is a random sample from the population of all hospitals in England is which the average mortality rate is 5% within 30 days of the surgery. If a particular hospital performed n procedures with a probability $p = .05$ of deaths following surgery we have n Bernoulli trials with mean number of deaths equal to np and standard deviation σ being the square root of $p(1 - p)$. The Central Limit Theorem applied to Bernoulli trials tells us that for n large enough, as will be the case, with 95% confidence the quantity S_n will lie within the interval np plus or minus $2\sigma/\sqrt{n}$. A 95% confidence interval for the unknown p is an interval estimate for the range of possible p values that is computed from the observed data $\hat{x} = Sn/n$. With $\sigma = \sqrt{p(1 - p)}$ the normal approximation gives

$$\hat{x} - 1.96\,\sigma/\sqrt{n} < p < \hat{x} + 1.96\,\sigma/\sqrt{n}$$

The confidence level is the proportion of confidence intervals that contain the true but unknown value of p and these vary since $\hat{x}$ is a random variable. This means that if the sampling is repeated many times the probability that the true value of p lies within at least one such interval is .95 but there is a probability of .05 that no such interval includes p. Thus any specific interval may or may not contain p.

For the hospital data $2\sigma/\sqrt{n} = .014$ when $n = 1000$ and so the mortality rate in this case is between .036 and .064 with 95% confidence while a larger hospital doing $n = 4000$ surgeries the mortality rate is now within a smaller interval of .043 and .057, closer to the overall average of .05. so, *as in the example of cancer mortality rates, small hospitals (doing less surgeries) are more likely to be listed among the most and least safe hospitals in England.* This not to say that there is in fact some other underlying reason for good and bad outcomes based on financial and managerial concerns and even the degree of skill of the surgical staff. These intrinsic factors may or may not overshadow the sheer luck associated with large variability in performance due to hospital size. Evidently the hospital with 4000 procedure has a correspondingly larger number of surgical deaths but the per-capita rate is nonetheless smaller due to the factor of $1/\sqrt{n}$. The data provides a single snapshot of surgical mortality at some particular time. Viewed at some later period the ratings of the different hospitals could shift up or down due to happenstance.

One sees this also in the performance of schools of different sizes, among similar examples. All this vaguely resembles the phenomenon of regression to the mean in which two variable quantities of some attribute are correlated and are viewed at sequential periods of time. Success in one period can be followed by less success in a subsequent period due to an intrusion of chance such as when a student who scores well on an exam does less well on a similar test later or a baseball player doing poorly in one or more games does better in subsequent games. What we have here are correlated normal variables in which a specific attribute such as performance skill is compromised by exogenous random factors. This differs from the variability of a single normally distributed variable such as surgical mortality in the hospital example.

The interplay between skill/talent and luck/chance, is a core idea that we encountered earlier in this book, and will do so again. Good expositions for these somewhat elusive circle of ideas, which began with Galton's book published in 1886, are Stigler's article [34] and Ellenberg's [35] in which a number of other statistical conundrums are discussed.

Chapter 10
The Wrong Conclusion

Medical screenings can be a source of well intentioned but seriously flawed judgements by doctors due to a cognitive failure in assessing conditional data. I provide a striking example here concerning a common procedure to screen for colon cancer in which, as will be seen, Bayes Theorem lurks in the background.

Two separate studies on the effectiveness of colonoscopy were published together in the July 2000 issue of the *New* England *Journal of Medicine* [54]. More than 5000 individuals underwent colonoscopies in the combined investigations with the goal of determining the extent to which polyps in the lower colon (those parts of the rectum and colon that are reachable by a sigmoidoscopy) were or were not indicators of advanced lesions in the upper colon (which can be reached only with a full colonoscopy).

The conclusions arrived at in both studies were consistent: a significant percentage of people with advanced lesions (adenomas) or cancerous polyps in the upper colon would not have had the disease detected had they relied solely on a sigmoidoscopy (51% in one study and 46% in the other). Therefore, in spite of the considerable cost, inconvenience, and a slight risk of procedural complications, the investigators recommended a full colonoscopy, a finding that was underscored in the NY Times in an article published in 2000 [55] from which we read" the theory that if no abnormalities were seen in the lower colon none were likely to be found higher up is contradicted (in the new NEJM study) which shows that it is not safe to assume the upper colon is healthy just because the lower third look normal." *But this is an erroneous conclusion and inaccurate reporting, as I wish to show.* I'll do this below.

From my discussions with several gastroenterologists it appears that these findings have altered their practice in the decade since the report from one that relied primarily on the simpler and less expensive sigmoidoscopy to the now routine recommendation of a full colonoscopy for the majority of their patients.

There is, however, another conclusion to be drawn from the same data that at first glance almost seems contradictory, namely, of all individuals showing no lesions in the lower colon (56%, more than half of all those examined) only a small percentage

E. Beltrami, *Digressions in Elementary Probability*, Quantitative Methods in the Humanities and Social Sciences, https://doi.org/10.1007/978-3-031-70051-4_10

will actually be found to have advanced upper colon disease (27 out of a thousand in one study and 15 out of a thousand in the other). This speaks in favor of a sigmoidoscopy alone.

This can be explained quite simply by looking at the bottom line of the data reported in studies I and II (I use the medical terms distal and proximal for lower and upper colon, respectively, in which distal includes the descending colon):

		Study I	Study II
A	total people screened	3121	1994
B	individuals with no distal polyps among those in A	1765	1564
C	individuals with advanced proximal lesions among those in A	128	50
D	individuals with advanced proximal lesions among those in B	48	23
E	individuals with no lower colon lesions among those in C	66	23

Observe that the probability of advanced upper colon lesions, given that there are no lower colon polyps, is approximately the ratio of D to B, namely 48/1765 = .027 in one case, and 23/1564 = .015 in the other.

On the other hand, the probability of no lesions n the lower colon, given there is advanced upper colon disease, is approximately the ratio of E to C, namely .51 in one case and .46 in the other.

Though these assertions sound superficially alike, they are not the same. Let U denote the event that there are advanced proximal lesions and L that there are no distal lesions. The fallacy lies in confusing the probability of L, given U with the probability of U, given L. What has happened is that the population of individuals with no distal lesions is quite large and only a relatively small number of these manifest serious lesions in the upper colon; by contrast, those with serious upper colon disease is a relatively small fraction of the total population but these include a fairly large percentage of people with no polyps in the lower colon. The underlying sample populations for the two statements are therefore quite different, a fact that is a source of some confusion in statistical reasoning in general and is another example of a base rate fallacy. The simple diagram in upper part of the Figure below captures the entire argument in a nutshell in which L is the set of people with no polyps in the lower colon, U is the set of individuals with lesions in the upper colon and the shaded area consists of patients for which both are true, namely the percentage of people in L with upper colon lesions and, at the same time, the percentage of those in U consisting of those with no lower colon polyps. *Two very different interpretations, easily confounded, because they are percentages relative to different base populations.*

From the point of view of a physician it makes sense to behave with caution and to quote the second statistic to a patient, recommending a full colonoscopy since "about 50 % of all cancerous polyps (in the upper colon) would not have been found from a sigmoidoscopy alone" This is entirely correct as a statement but it needs to be reconciled with what a patient (with no family history and trying to avoid the discomfort, risk, and considerable expense of a colonoscopy) states, in an equally correct manner, that "if a simple sigmoidoscopy shows no polyps then there is only a very small chance that my upper colon will exhibit any cancerous lesions". This conclusion is what would be relevant to most people.

Based on the data, both statements are true but when I tried to explain this to gastroenterologist he insisted, with some irritation, that the first statement is the only relevant one, disregarding the second. Moreover, some recent studies question the efficacy of colonoscopies as a tool for reducing the risk of cancer compared to sigmoidoscopy and they are no longer generally regarded as a panacea as was the case in earlier years (see [54, 55]).

To put this in a Bayesian context note that the prior probability of proximal disease is the ratio of B to A which in study I is 128/3121 = .041 and, in study II, the ratio is 50/1994 = .025, both modest in size. But, as we saw, if one now performs a sigmoidoscopy on an individual which reveals no distal polyps then, given this new information, the posterior probability of proximal disease is now even smaller, only .027 in the first instance and .015 in the second, which is about half in each case.

The lower part of the next figure illustrates a quite similar conclusion that can be verified by consulting the latest state-wide data on Covid-19 hospitalizations. A comparable percentage of vaccinated and unvaccinated individuals are hospitalized leading one to question the effectiveness of vaccination. However, the shaded area, representing those in the hospital, is but a small fraction of all vaccinated people, the set V, while being also a large fraction of the much smaller set U of unvaccinated individuals (at least is a state like New York according to my last perusal of published data). Focusing only on the shaded area without considering the size of the intersecting base populations leads to an unfounded conclusion. This is virtually a paraphrase of the colonoscopy example resulting in very similar figures (Fig. 10.1).

A mention of Covid-19 leads to a question of how accurate antibody tests are. Using data from the spring of 2020, also from New York at the beginning of the Pandemic, the *prevelance* of infections was 10%. In testing people for antibodies the *sensitivity* of the diagnostic test, namely there probability a positive test result T given an individual is infected, was 87.5% whereas the *specificity*, the probability of testing negative when a person is not infected, was 97.5%. Letting C be the event a person carries the antibodies we want to find the probability of C given T. Bayes Theorem is again the appropriate tool here. Note that the false positive rate, the probability of testing positive when ~C is true, is one minus the specificity, namely 2.5%. The prior odds of C are prob(C)/prob(~C) = 1/9 (probability of 10%) while prob(T || C)/prob(T || ~C) = 87.5/2.5 and so the posterior odds of C given T is 87.5/22.5 = 3.89; this gives a probability of 3.89/4.89 = .795. So, having tested positive, an individual is about 80% certain of having Covid-19 antibodies (see the N Y Times article [49]).

Well before colonoscopy there was a simple test for colon pathology, the fecal occult test. Though not very effective it is non-invasive and inexpensive. Here too the base rate fallacy intrudes to confound some physicians who use it as a diagnostic tool to screen for abnormalities as discussed by Gigerenzer [28]. I present it here

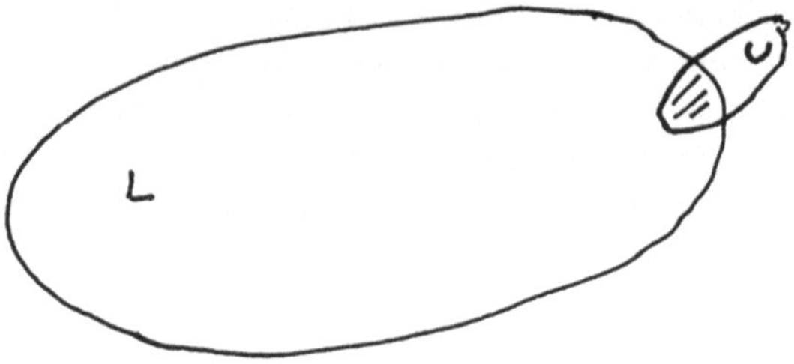

L is the set of people without polyps in lower colon; U is set of people with serious upper colon lesions; shaded area are people with both conditions

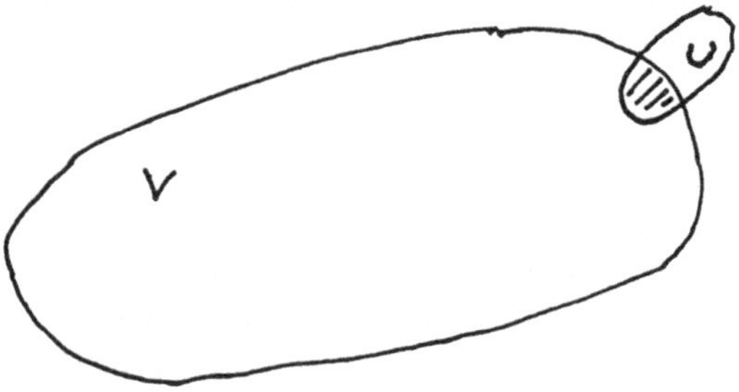

V are people who are Covid vaccinated; U are people who are Covid unvaccinated; shaded area are people in the hospital being treated for Covid complications

Fig. 10.1. *L* is the set of people without polyps in lower colon; *U* is set of people with serious upper colon lesions; shaded area are people with both conditions. *V* are people who are Covid vaccinated; *U* are people who are Covid unvaccinated; shaded area are people in the hospital being treated for Covid complications

briefly to give another slant on how conditional probabilities can be easily misinterpreted.

Let C denote the set of people within a given population that have colon cancer and that the probability of C, namely the fraction of individuals within the population having the disease, is .003. T will denote a positive hemoccult test and, since the test is not very sensitive, P(T || C) = .5. The probability of false positives, however, is only .03. that is, P(T || ~C) = .03. Suppose you test positive. A physician might be tempted to say “this test is very accurate (namely it has a small false positive rate) and it is therefore very unlikely that that someone without cancer would test positive.

But you did test positive and so it is very likely that you do have cancer.". *This an incorrect diagnosis, however, whenever the base rate for the disease is small as it is in this case.* I'll show this in a moment. What the patient wants to know is P(C || T) and not what the physician reports, namely P(T || ~C). These are not the same quantities and our medical practitioner confounds these probabilities. I want to show this using Bayes Theorem in odds form as discussed earlier in these notes.

The prior odds for colon cancer are P(C)/P(~C) = .003/.997. Moreover, the Bayes factor is P(T || C)/P(T. || ~C) = .50/.03 and so the posterior odds are .50/.03 *times* .003/.997 = .05. It follows that P(C || T) = .05/1.05 = .048 and so approximately one person in 20 who tests positive will have colon cancer. Not so scary as it first appeared. And the reason, in hindsight, is that since the disease prevalence, namely the base rate for cancer, is so small most people who test are cancer free. But this is a large fraction of the population and therefore even though the false positive rate is small a considerable number of people will still test positive even though they are disease free. To put this in different terms, out of ten thousand people, 30 have colon cancer and the rest do not. The fecal test is only half accurate and so only 15 of the ill individuals will test positive. Among the other roughly 10,000 people about 300 test positive and so the probability of having the cancer is roughly 15/315 or about .05 which accords with the Bayesean solution.

Chapter 11
More Fallacies

The way data is aggregated can lead to disturbingly different interpretations that can appear to be paradoxical. I will discuss two such in this section. The first is known as *Simpson's paradox* which has already come up in our previous discussion of "hot hands" in basketball (see, in this connection, reference [16]). Simpson, by the way, is the same Simpson that makes an appearance later when we discuss code-breaking at Bletchley Park during the Second World War.

Roughly speaking the conundrum arises when one finds that aggregated data leads to a conclusion that is at odds with what desegregated data seems to imply. Since we have been discussing baseball I'll begin with an example introduced by Ken Ross [57] involving the *batting averages* of two players, Derek Jeter and David Justice, in two consecutive years as tabulated below in terms of number of hits divided by the number of *at bats* in a season:

	1995	1996	combined
Jeter	12/48 = .250	183/582 = .314	195/630 = .310
Justice	104/411 = .253.	45/140 = .321	149/551 = .270

What one sees is that Justice is a better batter in the separate years but when the two seasons are combined it is Jeter who has the better batting average. What appears to have happened is that the disparity in the number of *at bats* for the two players in each season leads to empirical probabilities of getting a hit, namely batting averages, that are sufficiently lopsided to reverse what is found in the combined averages. Sample size is the culprit here much as base rate is in medical screening as we observed in an earlier section.

So who is the better player? Based on the more reliable combined data one ventures to say Jeter, at least in terms of the probability of getting a hit, but doubts linger.

Since we are discussing sports let me return to the question of "hot hands" in basketball as seen by Wardrop [16] in the fifth vignette. He compared the free throws of Larry Bird and Rick Robey, two star players of the Boston Celtics, during the

E. Beltrami, *Digressions in Elementary Probability*, Quantitative Methods in the Humanities and Social Sciences, https://doi.org/10.1007/978-3-031-70051-4_11

1980–81 and 1981–2 seasons. The following two tables record their hit and misses after a first successful hit.

	Larry Bird			Rick Robey		
	Second Throw			*Second Throw*		
	Hit.	Miss	Total	Hit	Miss.	Total
First Throw						
Hit	251.	34.	285	54.	37.	91
Miss	48.	5.	53	49.	31.	80
Total	299.	39.	338	103.	68.	171

Let p_{hit} and p_{miss} denote the proportion, the empirical probability, of a first throw hit or miss, respectively, being followed by a hit. For Bird, $p_{hit} = 251/285 = .881$ and $p_{miss} = 48/53 = .906$ whereas for Robey the numbers are .593 and .612. Neither table makes a conclusive case for a 'hot hand" since each player shot slightly better after a miss than a hit. However, when the tables of both players are combined into one we now observe a clearer display of this phenomenon as shown in the next table in which p_{hit} is now .811 and $p_{miss} = .729$, again illustrating Simpson's paradox as well as providing some support for the "hot hands" belief in the sense that there is in fact a significant probability of one or the other player getting a hit following a hit. This is to "hot hands" in the large, so to speak, namely as the game progresses. and argues in favor of this phenomenon being real as suggested by Miller and Sanjurjo in an earlier section.

	Second Throw		
First Throw	Hit.	Miss	Total
Hit	305	71	376
Miss	97	36	133
Total	402	107	509

For a last example I turn to data in Simpson's original paper [59] as paraphrased in the article [58]. It concerns a medical treatment on a sample of 52 individuals which shows one result for the whole sample but an different association when men and women are considered separately. The treated individuals are labelled by T and a control group (untreated) by C. The treatment can either be a success S or a failure F. There are 20 men and 32 women. The tableaux below summarizes the results:

	Full Population (52)		Men (20)		Women (32)	
	S	F	S	F	S	F
T	20	20	8	3	12	15
C	6	6	4	3	2	3

The treatment is ineffective at the level of the overall population but is more successful both for men and women separately (61% versus 57% for men and 44% versus 40% for women) than when there is no treatment. Here, then, is a quandary. Should the treatment be applied? If the gender of the patient is known then presumably yes but otherwise no which seems strange since a patient will still be either male or female even if their gender is unknown. One disparity in the data above is that though the treatment group is mainly female the control group is primarily male.

So the paradox can arise because of warped sample sizes or because of gender disparity. Other lurking variables could also lead to a reverse conclusion when aggregated data is disaggregated. The paradox is more common than may be imagined at first.

A second example goes by the name of *Berkson's paradox* and it once again illustrates, as we have seen several times, the befuddlement that can arise when using conditional probabilities. It's a statistical quirk, sometimes referred to as a *selection bias*, that shows a negative correlation between two traits when in fact they are unrelated. This occurs when one selects a subset of the entire population that ignores those individuals that have neither trait.

The original paper [61] from 1946 was in the context of two independent illnesses in the general population, diabetes and bladder inflammation, that become negatively correlated when one considers a subset of these individuals having one or the other ailment, the subset being people that are hospital in-patients. A hospital patient without diabetes is more likely to have bladder inflammation than a member of the general population since the patient must have had some non-diabetes illness to enter the hospital in the first place.

To put this in a more specific setting suppose that some people in a community have one or the other or both of two medical ailments, say A and B or, perhaps, one considers students in a high school who have a good GPA or good SAT scores, possibly both (again label these attributes as A and B). I assume that not all individuals in the community have the two ailments nor, for that matter, do all of the students have good GPA and or SAT numbers. Those people without one ailment or the other would not appear in a hospital survey of A or B nor would students without the requisite grades be found among the college applicants. Thus, $P(A \cup B) < 1$.

I now show that regardless of P(A) and P(B) within the community or high school it will be true that $P(A) < P(A \parallel A \cup B)$ with an identical result when we interchange A and B. To see how this comes about assume, as already mentioned, that $P(A \cup B) < 1$.

Then $P(A) \leq P(A \cup B) < 1$ since $A \cup B$ includes A. But $P(A) = P(A \cap (A \cup B))$ and, by the definition of conditional probability,

$$P(A) < P(A)/P(A \cup B) = P(A \cap (A \cup B))/P(A \cup B) = P(A \parallel A \cup B) \qquad (11.1)$$

This means that within the subset $A \cup B$, the events A and B can be *negatively correlated even if they are uncorrelated or even positively correlated in the general*

Fig. 11.1 Conditional probability

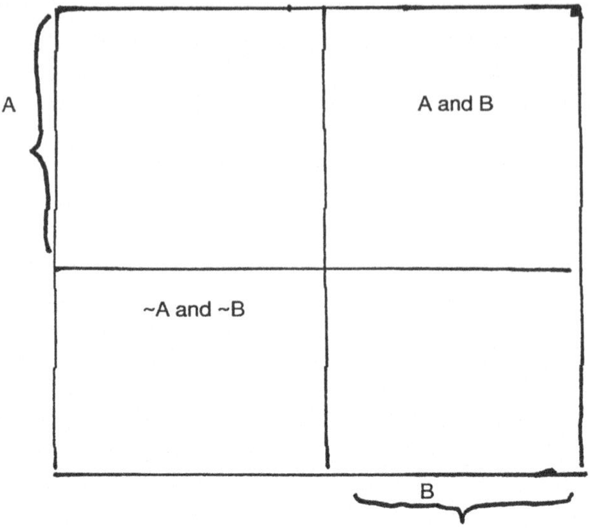

population. I'll illustrate this by looking an example to be found in a paper by Miller and Sanjurjo [6] in which the High School students considered just above have a .5 probability of doing well in their GPA and, independently, also .5 chance of scoring well in the SATs so that PA ∩ B) = 1/4. These attributes, A and or B, together account for 3/4 of the school enrollment as can be see from the figure below (~ means negation, namely the attribute does't hold). Thus the probability of A and or B ≡ P(A ∪ B) = 3/4 and it therefore follows from (11.1) that P(A || A ∪ B) = 1/2 divided by 3/4 = 2/3. So although 1/3 of the students in A ∪ B have both attributes two thirds of them have either good GPA or good SAT *but not both*. This implies a negative correlation between the attributes (Fig. 11.1).

The same conclusion can be reached using Baye's Theorem in odds form very much as we did in chapter 6 in connection with problems of restricted choice. I need to expand the narrative somewhat by adding that the group having A or B applies to college whereas 1/4 of the students having neither A or B do not apply. Let E be the event of A and ~ B while E* is A and B and let F designate the event that a student applies to college using their best academic attribute. The prior odds of E are one because of independence of A and B. However, though P(F || E) = 1 the probability P, (F || E*) is only 1/2 since either A or B could equally have been chosen as best attribute. Therefore the posterior odds of E given F is 2:1 leading to a probability of 2/3 that a student has A and ~ B, the same conclusion that was reached above using conditional probability.

The two disease problem can be also be handled by Baye's Theorem in a similar manner. Letting A and B be the independent events of having one disease or the other (say high blood pressure and diabetes) we let E and E* have the same meaning as before while F is the event that an individual enters a hospital because of one ailment or the other. This is the equivalent of applying to college in the previous example. Then an identical analysis shows that the probability of having A but not B as recorded in the hospital is 2/3 assuming the probabilities of A and B are separately 1/2.

A nice discussion of this hospital paradox using different and more realistic data can be found in Ellenberg [35] with diabetes and high blood pressure as the two ailments. Quoting from Ellenberg we read that "If you're in the hospital, you're there for a reason. If you are not diabetic that makes more likely the reason is high blood pressure. So what at first looks like a causal relation between blood pressure and diabetes is really just a statistical phantom."

Chapter 12
Clinical Trials

A typical setting for testing a hypothesis in clinical trials would be a comparison of two clot-busting drugs on mortality following cardiac surgery or whether a drug is more effective than a placebo (aspirin, say, in preventing heart attacks within a five-year span). In each case two groups of individuals are recruited having similar medical profiles and then randomly assigned to one protocol or other. We want to test whether any observed difference between the outcomes can be considered statistically significant. Typically the subjects are not informed as to which group they belong. One starts with is a fervent hope that there is a meaningful difference between the two outcomes. To do this one begins by invoking a *null hypothesis* H_0: any difference one observes is due solely to chance. The null hypothesis is the straw man of statistical testing in which one plays the devil's advocate in the hope that it can be refuted. This is analogous to proving a theorem in mathematics by assuming the result is false and then showing that this leads to a contradiction implying that the result must in turn have been true all along. If H_0 is true there will still be detectable differences due to sampling errors but can we say, with some degree of certainty, that the difference goes beyond experimental error and can be attributed to some real effect?

Note that if we do not reject H_0 this does not mean that there isn't a real difference between drug and placebo but simply that the experimental trial did not prove there is a difference at the given level of confidence. On the other hand, rejecting H_0 means either that the null hypothesis is not true or that it is true and the value of a test result is a fluke.

I want to illustrate the effectiveness of vitamin C in curbing some cold symptoms by comparing it to a placebo. The details are in the paper by Berger and Berry [52].

A total of 17 paired samples of subjects are randomly chosen, each pair as similar as possible, and one is given Vitamin C and the other a placebo P as a double blind choice. After 24 h we check to see which subject within a pair exhibits some specific level of relief from the cold symptoms. The null hypothesis H_o is that Vitamin C has no significant effect over the placebo and that any relief in symptoms is due solely to random fluctuations that are always part of an experiment. That is, each preference

E. Beltrami, *Digressions in Elementary Probability*, Quantitative Methods in the Humanities and Social Sciences, https://doi.org/10.1007/978-3-031-70051-4_12

for C over P and, vice-versa, for P over C in each pair is like tossing a fair coin (probability of ½ for either outcome) and seeing if we get a head or tail. It turns out that we find that C is better in 13 pairs and P is better in the remaining 4 pairs. Is this significant? We are comparing the number of trials in which C is better that P or P is better than C to obtain evidence against the null hypothesis that neither is better than the other. If C is much larger than P then we are inclined to think that C is beneficial whereas a larger number of successes for P would suggest that C is detrimental; either result casts doubt on H_o. what we have, therefore, is a two-sided test.

Since we are dealing with something akin to 17 coin tosses one can ask for the probability of 13 heads (C is better), namely prob.(13 successes in 17 trials) for the binomial distribution. In Figure 12.2 we show the plot of the binomial distribution for 17 trials, namely for $k = 0, 1, 2, \ldots, 17$ successes (preference for C). The probability of the actual outcome of the experiment is the height of the bar of unit width over 13, which is only .0182. What we want is the probability of results *equal to or more extreme* than what we obtained experimentally. In the figure this constitutes the area of all the bars from 13 or higher or 4 or less (4 or less heads in the 17 throws is just as extreme as 13 or more). When we add up all these areas the total probability turns out to be .049 which is less than the usual cutoff value of .05 for significance. This provides a 95% confidence level for the rejection of the null hypothesis. That is, if H_o were correct we would have expected that 95% of the trials would have led C being better in not more than 12 or less than 5 pairs. But now we see a problem. The actual contradiction of H_o involved not only the observed 13 preferences for C but also the unobserved possibilities of 0, 1, 2, 3, 4 and 14, 15, 16, 17 as outcomes, data for which we have *not seen*.

What is the case here is that instead of asking for the probability of an experimental outcome given a null hypothesis, which is the usual protocol in medical and pharmaceutical trials, what is needed is the quite different conditional probability of the null hypothesis given the experimental outcome. This was carried out in [53] using the same Vitamin C trial discussed above and it shows that the smallest evidence against H_o is actually .21 which is 4 times higher that .05 and this is considerably milder evidence for rejecting H_o.

Another problem with the usual approach to hypothesis testing is that it is difficult to assess just what was in the mind of the experimenter. The 13 positive outcomes for C could also have resulted from an entirely different protocol for which, instead of using 17 matched pairs of subjects, the investigator had decided to stop the experiments the moment that at least 4 pairs were obtained for which C is better and 4 for P; when that first happens the experiment is ended. But now suppose that the fourth P occurred on the 17*th* trial and that the results are the same as before: 13 for C and 4 for P. So H_o is again rejected but for different reasons and one doesn't know which is which unless the investigator tells us. Did he or she stop because no additional matched pairs could be found or because funding ran out or, perhaps, because the investigator felt that the evidence gathered was sufficient?

I now consider a variant of the above experiment which is applicable to early stage trials involving 17 single patients rather than matched pairs. I use a modification of the discussion in [53] in which the goal is get an early estimate of the efficacy of the drug. When 5 individuals react positively the trials are ended and the drug is deemed successful enough for a full-blown trial. The first figure below exhibits the negative binomial distribution that is applicable here. What this entails is a random variable X that indicates the number of independent trials necessary to obtain k successes. The probability that X equals n, for $n \geq k$, is obtained by noting that for the k th success to occur on the n trial there must have been $k - 1$ successes in the previous $n - 1$ trials and this can be done in $C(n - 1, k - 1)$ ways. This event is then multiplied by the probability p of the independent event of success on the n th trial to obtain

$$\text{Prob}(X = n) = C(n - 1, k - 1)p^k(1 - p)^{n-k} \tag{12.1}$$

Note from the second figure below that the likelihood of k successes on any one trial is not very large but there is a better than even odds of this happening if the trials continue up to their maximum whenever $p = 1/2$.

Though (12.1) is superficially similar to the expression for the binomial distribution what in fact varies in (12.1) is n for a fixed k whereas in the case of the binomial n is now fixed and k varies to obtain the probability of k successes in n trials. In the Appendix is a MATLAB code for the second figure below called *negbinomial.m*

I want to comment briefly on the standard protocol for hypothesis testing in clinical and pharmaceutical trials. Typically a *statistic* is invoked, such as the difference between two sample means from a normal density, a numerical measure of variation that can discriminate between sampling error and a real effect. Suppose the experimental data gives a value of S for this statistic. The null hypothesis H_0 is that the value of S is due solely to random fluctuations in the data. But suppose that the probability of finding S under the null hypothesis is very small, usually set at 5% or less. Then the result is deemed to be statistically significant. That is, if chance alone produces the outcome we observe in less than one in twenty trials (probability .05) we reject H_0 and accept the alternative that something real is going on. As already mentioned, a rejection of H_0 at a significance level of .05 does not mean that there is only a 5% chance the result is a fluke nor does it imply that the experimental results are significant with probability .05. It simply tells us that the null hypothesis is untenable at the 5% confidence level using the particular test statistic. This, of course, is simply a variant of the idea of a 95% confidence interval as discussed in an earlier vignette and in the Tutorial. The area under a normal curve that has zero mean and unit variance lying between plus and minus 1.96 is .95 and the area outside this interval, at the tail ends, is then .05.

Rejecting a null hypothesis as just described has been criticized, as also mentioned earlier, for being too extreme tending to undervalue H_0. A Bayesean approach is generally more robust in this regard. There are many pitfalls in hypothesis testing in medicine and pharmacology. A good review can be found in the article [70] by Lehrer (Figs. 12.1 and 12.2).

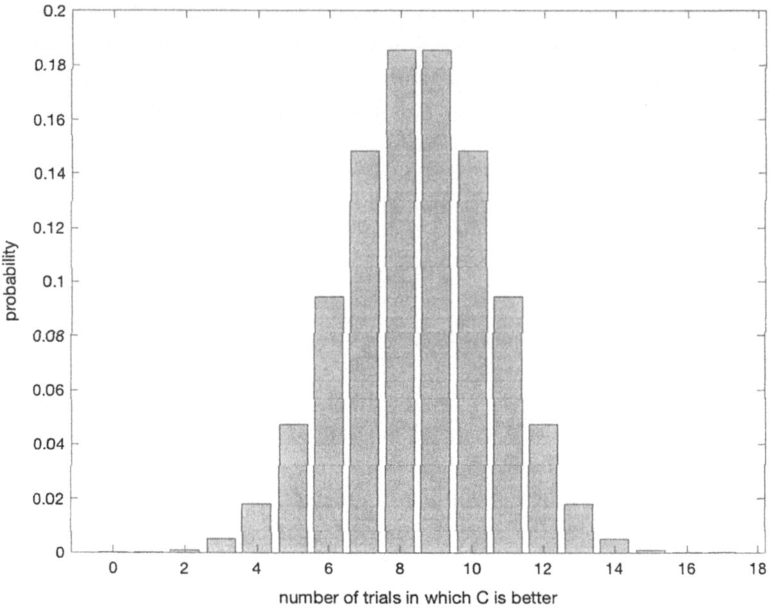

Fig. 12.1 Number of trials

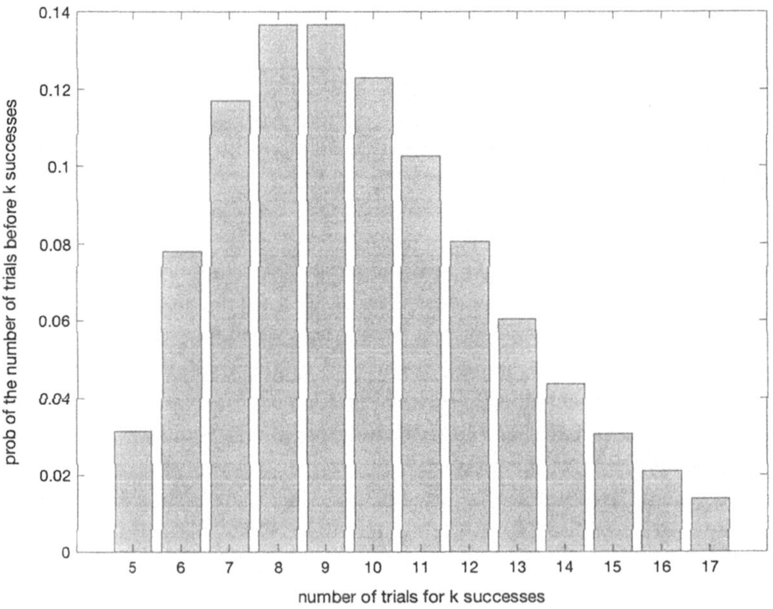

Fig. 12.2 Testing in clinical and pharmaceutical trials

Chapter 13
Hot Hands, Again

I return to the "hot hands" controversy by revisiting an earlier paper by Wardrop [16] that looked at basketball data for players Robey and Bird. The relevant data can be found in the earlier section called "More Fallacies" but for convenience the table of collapsed scores is reproduced below. My goal is to use the hypothesis testing procedure reviewed in the previous vignette to add something different to the biased estimation analysis of Miller and Sanjurjo of Section 4, again indicating a validation of sorts for players being "in the zone." I'll be testing the significance of second throws in terms of a successful hit being followed by another "make" using the *combined scores* of just the two players. A similar analysis exhibits no statistical significance when the players are considered separately.

First Throw	*Second Throw*		
	Hit.	Miss	Total
Hit	305	71	376
Miss	97	36	133
Total	402	107	509

To begin I'll discuss a test for the independence or dependence of a second throw hit following a first throw hit. Think of all second throws as balls in an urn containing m white balls each designating a hit and N – m green balls (a miss). Set up a null hypothesis that the hits in a second attempt following a hit is a random event devoid of any possibility of a player being "hot". There are, say, N first throws and these are followed by a hit or miss. If what happens in this second attempt is independent of a first success let X be a random variable giving the number of white balls picked at random from a sample of size n *without replacement (once you pick a throw and see if its a success or failure you can't repeat it)* from the urn following a successful first hit. There are C(n, k) ways of picking k white balls at random from the m white balls available and for each of these choices there are C(N – m, n – k) possible ways of then picking n – k green balls from the remaining n – k random draws. Moreover there are C(N, n) ways of a random selection of n balls from a total of N and so the

E. Beltrami, *Digressions in Elementary Probability*, Quantitative Methods in the Humanities and Social Sciences, https://doi.org/10.1007/978-3-031-70051-4_13

probability that X = k, where k can vary from 0 to n, is given by what is known as the hypergeometric distribution, a variant of the binomial density in which sampling is done without replacement:

$$P(X = k) = C(m,\ k)C(N - m,\ n - k)/C(N,\ n)$$

Thus P(X = k) is the number of hits in the second throw following a successful first throw. The average of X turns out to be nm/N and its variance is $Mm(N - m)(N - n)/(N - 1)N^2$, as I'll establish in an Addendum to this section. I will use these quantities shortly.

The above table combines the scores of Bird and Robey the column sums being m and N – m respectively with N being the total number of throws (hits and misses), n is the first row sum (sample of size n) and the second row sum is N – n.

A glance at the above table shows that there were H = 305 hits following a hit in a sample of n = 376 throws. The total number of hits following all the first throws is m = 402 and there are 107 misses for a total of N = 509 altogether. Using this data we can approximate the mean and variance of H and find E(H) = mn/N = 376 × 402/509 = 292.96. Moreover the variance of H becomes 16.34 with a square root of 4.04. The statistic Z = [H – E(H)]/ √ Var(H) is approximately normal with mean zero and variance one and has a value of 8.04/4.04 = 1.99 which lies just outside the 95% confidence interval (which would be Z equal to plus or minus 1.96) and so lies within the tail of the normal density. This corresponds to a probability of obtaining Z under the null hypothesis being less than .05. This encourages one to think that the null hypothesis can be rejected and that there is in fact a significant probability of one or the other player getting a hit following a hit, a "hot hand" in the large, so to speak, namely as the game progresses, a manifestation of "streak momentum".

In [16] Wardrop extended this analysis to nine members of the Boston Celtics to obtain an even more convincing validation of player momentum.

As an aside, the probability of k successes in the hypergeometric density can be computed by a single MATLAB instruction *hygepdf* that depends on k, m, n, N but we have no need for this here.

Addendum

Addendum: Mean and Variance of the Hypergeometric Distribution

We want to compute the expected value of H. Begin by noting that the probability of randomly picking a white ball in an urn of N balls when n are selected is m/N since the k*th* draw is equally likely to be any of the N balls in the urn with m of them being white balls ("success"). Said otherwise, the probability that a randomly picked ball in the urn is white is m/N. It does not depend on the order of the draw but only on the number of white and green balls, a property called *exchangeability*. I'll establish this in the special case of two white and one green ball in the Addendum that follows. This will suffice to make the general case credible. And therefore the random variable X_j which equals one if the *jth* draw is a success and zero otherwise will have an expected value of m/N. With n draws the expected value X is the sum of the individual averages X_j summed from 1 to n giving

$$E(H) = nm/N.$$

The variance is a bit more tricky and I'll follow the computation in Ross [13]. As in the previous paragraph let $X = \sum X_j$ summed over j from 1 to n. The $E(X^2) = E[(\sum X_j)(\sum X_i)] = E[\sum X_j(X_j + \sum X_i)]$ where the last sum is over all $i \neq j$. Then $E(X^2) = E\left[\sum X_j^2 + \sum\sum X_j X_i\right] = E\left(\sum X_j^2\right) + \sum\sum E(X_j X_i)$ where, again the sum over i excludes j. Since the X variables are zero or one then $E\left(\sum X_j^2\right) = E(X_j)$ and $E\left(\sum X_j^2\right) = mn/N$. By the same token $X_j X_i = 1$ if and only if both variables equal one and the product is zero otherwise. Clearly $E(X_j X_i) = P$ (draws j and i are successes).

Using conditional probabilities, $P(X_j = 1, X_i = 1) = P(X_j = 1 \| X_i = 1)P(X_j = 1) = m/N(m - 1)/(N - 1)$ because conditional on a first success there are now a total of N − 1 balls and m − 1 white balls. The sum over j has n terms and over i just n − 1 terms and so $Var(X) = E(X^2) - E(X)^2 = mn/N + n(n - 1)m/nx(m - 1)/(N - 1)$ and this simplifies, after some further manipulations, to $Var(X) = mn(N - m)(N - n)/[N^2(N - 1)]$.

Addendum: Exchangeability

Random variables $X_1, X_2, \ldots X_N$ are exchangeable if the joint density $p(x_1, x_2, \ldots, x_N) = P(X_1 = x_1, X_2 = x_2, \ldots, X_N = x_N)$ remains the same when the values $x_1, x_2, \ldots, x_N$ are permuted. The classic example is an urn with m white balls and N − m green ones. Let $X_k = 1$ if the kth selection from the urn is white and $X_k = 0$ otherwise. Each withdrawal is equally likely to be any one of the remaining balls. Let consist of m ones and N − m zeros. To see this in a specific case consider × = (1, 0, 1). This indicates that the first selection from the urn consisting of m = 2 white and N − m = 1 green balls has probability 2/3 since the first pick is white. Conditional on this event the next has probability of 1/2 since there is only one white and one green and we pick green. The urn now has only one white ball and, conditional on the previous selection, it is picked with probability 1 and so p(1, 0, 1) = 2/3 × 1/2 × 1/1 = 1/3. In a similar manner p(1, 1, 0) = p(0, 1, 1) = 1/3. It follows that $P(X_k = 1)$

is the same for $k = 1, 2, 3$ and we have exchangeability. Since $X_k = 1$ occurs in exactly two positions of the three vectors $x = (x_1, x_2, x_3)$ then $P(X_k = 1)$ is 2/3 for each k whereas $X_k = 0$ has probability 1/3 because it occurs only once for each k. In the more general case the same reasoning applies and so the probability of a white ball being chosen is m/N regardless of when the selection is made.

Chapter 14
Margin of Error

A group of individuals are randomly sampled from a given population and are asked their preference in terms of two political candidates. Since a random sample of preferences is subject to some level of uncertainty we want to assess whether any observed differences between these outcomes can be considered statistically significant. The margin of error for each candidate separately (which I'll describe more fully below) can overlap in which case there is a non-zero probability that they are in a draw and we need to compute this probability. The assumption throughout is that the samples are truly random with no compromising bias. That, of course, is a complicated issue about sampling protocols that I do not address here.

I'll begin by looking at the simpler case of the error involved with just a single candidate. Assume that n individuals are polled in a random sample and asked if they prefer candidate A. The true but unknown preference is p and the poll gives us a sample mean $x^\wedge = 1/n \sum x_A(k)$, namely a proportion in which the indicator variable $x_A(k) = 1$ if the k *th* sampled person prefers (votes for) A and is 0 otherwise. The sum is from 1 to n. Note that $\sum x_A(k)$ has a binomial distribution in which the Bernoulli random variables $x_A(k)$ have mean p and variance $p(1 - p)$.

Although I discussed this in an earlier section it may be helpful to repeat that for n large enough the distribution of sample means have an approximate normal distribution with mean p ($x^\wedge$ is an unbiased estimator of the true mean p) and a variance that in our binomial case is $p(1 - p)/n$. A 95% confidence interval for the unknown p is an interval estimate for the range of possible p values that is computed from the observed data $x^\wedge$. With $s = \sqrt{p(1 - p)}$ the normal approximation gives

$$x^\wedge - 1.96\ s/\sqrt{n} < p < x^\wedge + 1.96\ s/\sqrt{n} \tag{14.1}$$

The standard deviation s has a maximum of 1/2 and so (14.1) can be approximated by

E. Beltrami, *Digressions in Elementary Probability*, Quantitative Methods in the Humanities and Social Sciences, https://doi.org/10.1007/978-3-031-70051-4_14

$$x^\wedge - 1/\surd n < p < x^\wedge + 1/\surd n \tag{14.2}$$

Interval (14.2) is called the *margin of error*. The confidence level is the proportion of confidence intervals (14.2) that contain the true but unknown value of p and these vary since $x^\wedge$ is a random variable. This means that if the sampling is repeated many times the probability that the true value of p lies within at least one such interval is .95 but there is a probability of .05 that no such interval includes p. Thus any specific interval may or may not contain p.

This is illustrated by a simple example. Suppose $n = 1000$ and that 520 people prefer candidate A so that $x^\wedge = .52$. Since $1/\surd 1000$ is roughly .031 then plus and minus this number is the margin of error and we conclude that 52% of the people prefer A plus or minus 3.1 percentage points, namely the true probability of voter preference for A is $.52 \pm .031 = .479$ to .551 or, put another way, between 48 percent to 55 percent of the people prefer A with the given level of confidence.

If there are several candidates then a vote for one of them is a vote against the others and so the several probabilities that one needs to consider are (negatively) correlated. Pick any two of the candidates, call them A and B, and let $x^\wedge_A = 1/n \sum x_A(k)$ and $x^\wedge_B = 1/n \sum x_B(k)$ both summed to n. The sample mean $x^\wedge_A$ is the estimate of x_A and $x^\wedge_B$ the estimate for x_B. The margin of error for a single proportion can be extended to the case of two candidates A and B in the same poll. However the error intervals for each of them separately could well overlap in which case there is a chance that the voters have equal preference for each of them, a "dead heat". Therefore it is best to compute the margin of error for the difference between them. An example is 441 voters who preferred McCain at 30% over Giuliani at 26%, with a lead of 4 percentage points in a poll taken in 2007 as discussed by Franklin [29]. Was this significant? As I show below it was not .

In order to derive an expression for the difference in candidate preference I'll briefly review some facts about the covariance of two random variables (see a discussion in the Tutorial) which applied to $x^\wedge_A$ and $x^\wedge_B$ gives $\mathrm{Cov}(x^\wedge_A, x^\wedge_B) = \sum\sum \mathrm{Cov}(x_A(j), x_B(k))/n^2$, each sum up to n. If j is not equal to k the corresponding terms in the double sum are zero since each person votes independently for one candidate or the other and so one gets a single sum $\sum \mathrm{Cov}(x_A(j), x_B(j))/n^2$. Now $\mathrm{Cov}(x_A(j), x_B(j)) = E(x_A(j)\, x_B(j)) - E(x_A(j))E(x_B(j))$.

Since person j votes for A if and only if she doesn't vote for B, namely $x_A(j) = 1$ if and only if $x_B(j) = 0$, then $E(x_A(j)x_B(j)) = 0$ leaving $-\, x_A x_B$. It now follows that

$$\mathrm{Cov}(x^\wedge_A, x^\wedge_B) = -x_A\, x_B/n \tag{14.3}$$

To continue recall that $\mathrm{Cov}[(x^\wedge_A - x^\wedge_B), (x^\wedge_A - x^\wedge_B)] = \mathrm{Var}(x^\wedge_A - x^\wedge_B)$. But, as we saw, the covariance of two variables Y, Z equals $E(YZ) - E(Y)E(Z)$ and if we apply this to $Y = Z = x^\wedge_A - x^\wedge_B$ and then expand out the terms what results from this is

$$\mathrm{Var}(x^\wedge_A - x^\wedge_B) = \mathrm{Var}x^\wedge_A + \mathrm{Var}x^\wedge_B - 2\mathrm{Cov}(x^\wedge_A, x^\wedge_B)$$

Then, using (14.3), we find Var($x^\wedge{}_A$ − $x^\wedge{}_B$) = [x_A(1 − x_A) + x_B(1 − x_B) + x_Ax_B]/n and so, finally,

$$\mathrm{Var}(x^\wedge{}_A - x^\wedge{}_B) = \left[x_A + x_B - (x_A - x_B)^2\right]/n \tag{14.4}$$

Returning now to the McCain-Giuliani contest, the margin of error of (McCain—Giuliani) is, using (14.4), approximately 1.96 times the square root of [.30 + .26 – (.30 − .26)^2]/441, roughly .07 or 7%. So the difference of 4 percentage points in McCain's lead is not outside the margin of error and therefore is not statistically significant.

Finally, as discussed by Franklin [29], one can consider the difference in two independent polls and here the difference is computed using 1.96 times the square root of [x ^ (1 − x^)/n + y ^ (1 − y^)/m] in which the sample sizes are n and m respectively and x^, y^ are the sample means for each of the independent polls. Because of independence the covariance term in (4) is zero otherwise the same strategy for estimating a margin of error continues to apply. An example would be comparing the approval rating of the president in two independent polls taken three months apart.

Chapter 15
What Are the Odds of That?

The spooky quality of coincidences rarely fails to fascinate and confound people who experience them. What I hope to show is that many, perhaps most, coincidences are less amazing than they first appear to be.

Sample spaces in probability contain all possible events and not just those that catch our attention but we tend to focus on those meaningful to us. In effect, when a coincidence appears that we happen to notice, *what is being ignored is the larger number of other events that also lead to striking coincidences but that we failed to detect.* The source of wonder in an uncanny coincidence is our selectivity in reporting those events that catch our fancy. To quote Cohen and Stewart [31]: "The human brain just can't resist looking for patterns, and seizes on certain events it considers significant, whether or not they really are. And in so doing, it ignores all the neighboring events that would help it judge how likely or unlikely the perceived coincidence really is."

If you toss two identical balanced die (a cube with sides numbered from one to six) the sample space consists of 36 possible equally likely outcomes for the number of dots that appear on each die. We ask for the probability of getting the pair (3, 3). There is just one possibility here and so the probability of this happening is 1/36. However, if you ask for the probability of the same number coming up on each die, *any* number from one to six, there are now six possibilities to consider: (1, 1), (2, 2), ..., (6, 6) and so the probability of this event is now $6/36 = 1/6$. Asking for a particular coincidence is quite different from asking about any coincidence.

To provide a guidepost to the phenomena of coincidence I consider first the familiar *birthday problem* in which a group of k people are assembled quite arbitrarily and one inquires what the probability is that at least two of these individuals have the same birth date. A modest assumption is made here, one that is not quite true, that all birthdays are equally likely to occur on any day of the year and that a year consists of 365 days. Later I extend this question to ask a more subtle question, namely what the probability is that at least two individuals have a birthday no more that one day apart (near coincidence).

E. Beltrami, *Digressions in Elementary Probability*, Quantitative Methods in the Humanities and Social Sciences, https://doi.org/10.1007/978-3-031-70051-4_15

The probability that at least two of the k individuals share the same birthday is one minus the probability that all k persons have different birthdates. To compute this observe first that each of the k randomly chosen individuals can declare any one of the n dates as a birthday. Some of these may coincide, some not. So the sample space of all possible birthdays is obtained by taking a sample of size k with replacement from the total n and therefore the sample space has size n^k. Now consider the event that all k birthdates are distinct. This means that the first person can choose any one of $n = 365$ possible dates leaving the next person 364 choices while the third individual now has only 363 options, and so forth, until the last person has $n - k + 1$ possibilities, Therefore the probability of k distinct birthdays is $n (n - 1) (n - 2) \ldots (n - k + 1)$ divided by n^k which is the same as $k!\ C(n, k) = n!/(n - k)!$ and so the probability we want is $1 - n!/(n - k)!$ which can be expressed as

$$1 - \prod_{j=0}^{k-1} (1-j/365)$$

When k is as small as 23 this expression is greater than half and so there is a better than even chance that two or more of these 23 individuals will report the same birthdate! On the several occasions in which I've tried this experiment in a classroom of about 30 students only once did I fail to get an agreement on birth dates from two or more people in the class. The surprise here is that most students believe that it would require a much larger population of individuals to achieve such concordance.

The surprise happens because some people hearing it for the first time are responding to the wrong question, one that sounds superficially like the birthday problem, namely "what is the probability that someone else has the same birthday as mine?". The real issue is whether any two people in a room have the same birthday and there are many more possibilities for this to occur; we are fooled into thinking of the coincidence as something that happens between us and someone else rather than between any two randomly chosen individuals.

For the sake of demonstrating how different the probability that at least one of n randomly chosen people has the same birthday as myself note that 364/365 is the probability that some particular person has a birthday different from mine and, therefore, by independence, $(364/365)^k$ is the probability that none of the k people has the same birthday as me. Consequently, the probability that at least one person does in fact have the same birthdate as mine is $1 - (364/365)^k$. It takes $k = 253$ people before obtaining even odds of a match, and this number fits in better with the intuition of most individuals who first encounter the birthday problem.

In the case of near coincidence, when it happens to us, is often seen just as unsettling but the odds of this happening are even greater as I noted. When one of these people is yourself, the coincidence is striking but, of course, there are good odds it will happen to someone. When it does you either don't see it or you don't care. The phenomenon of my coincidence versus your coincidence has been examined through a set of psychological experiments which suggest that there is a real difference in perception of how surprising an event appears depending on who it happens to, as discussed by Falk [32]. In a national lottery, for example, in which a draw is six numbers out of the first 49 there are C(49, 6) = 13, 983, 816 possible outcomes and a winning draw is a long shot for you. But, out of the millions of people who play the lottery, many with multiple tickets, someone wins and if your ticket differs from the winning one by just one digit this may seem amazing but, in fact, many people have tickets that differ from each other by a single number but because they don't involve the winning draw no one notices.

This leads to the computation of the probability that k people have birthdays that differ by at most one day, an extension of the previous birthday problem. I establish this probability much as before as one minus the probability P(n, k) that each of the k people have birthdates that differ by at least two. Again n is 365.

The following proof is based on a paper of Abramson and Moser [30] and can be omitted since it is somewhat involved. What it shows, and it verifies our intuition, is that *the likelihood of a close coincidence may be more common than is generally thought.*

The derivation assumes that the first and last days of the year are consecutive, so that Dec 31 is followed by Jan 1. The birth dates for a random selection of k individuals are denoted by $1 \leq x_1 < x_2 < \ldots < x_k \leq n$ and we require that

$$2 \leq x_{j+1} - x_j \text{ and } x_k - x_1 \leq n-2, \text{ for } j = 1, 2, \ldots, k$$

The last condition assures us that if x_1 is, for example, equal to January 4, then perchance when x_k actually equals $x_1 + n - 2$ this would correspond to January 2 and so the gap of two between birthdates is preserved. In effect the k dates are taken modulo n.

The need for a somewhat different argument here than the one used earlier is the awkward fact that dates are separated by at least two. To convert this selection of dates into an equivalent set of dates that are separated by at most one day let's define new variables by $y_j = x_j - (j-1)$. Then $y_1 = x_1$ and $1 \leq y_1 < y_2 < \ldots < y_k \leq n-(k-1)$. It is immediate that $y_{j+1} - y_j \geq 1$ and that

$$y_k \leq n-(k-1) \tag{15.1}$$

Moreover, since $x_k \leq x_1 + n - 2 = y_1 + (n - 2)$ it follows that $y_k = x_k - (k - 1) \leq x_1 + n - 2 - (k - 1)$ and, since $y_1 = x_1$, we have

$$y_k \leq y_1 + n - k - 1. \tag{15.2}$$

There are two cases to consider: (i) $y_1 \geq 2$ and (ii) $y_1 < 2$, namely $y_1 = 1$. In the first case inequality (15.1) is smaller that inequality (15.2) and so (15.1) is the upper bound for y_k. Therefore $2 \leq y_1 \leq y_2 \leq \ldots \leq y_k \leq n - k + 1$ and so k birthdays can be chosen from $n = k + 1 - 1 = n - k$ available dates. This can be done in C(n – k, k) different ways. In the second case in which $y_1 = 1$ we obtain, using (15.2),

$$1 = y_1 \le y_2 \le \ldots \le y_k \le y_1 + n - k - 1$$

Since the first date is fixed there remains k – 1 possible birthdates to choose among the n – k – 1 available dates. This can be done in C(n – k – 1, k – 1) ways. Since C(n – k, k) = (n – k)/k times C(n – k – 1, k – 1) we can add cases i and ii to get n/k times C (n – k – 1, k – 1) possible ways for k people to select their differing birthdays. These individuals can be assigned to the different dates in k! ways (namely k! permutations of k people). Also, as before, the sample space of all possible birthdates, whether coinciding or not, is n^k. So, finally,

P(n, k) = k ! /n^k times n/k times C(n – k – 1, k – 1) which equals
(k – 1) ! /n^{k-1} times C(n – k – 1, k – 1) and the probability we want is

1 – P(n, k) = 1 – Π(1 – j/365) with the sum over j from k + 1 to 2 k as can be easily ascertained. Using this expression one finds that now there is a better than even odds of two individuals out of 14 having the same birthday or a birthday one day apart.

This is a good place to insert an intriguing problem, also involving coincidences, that requires simpler probability ideas for its resolution. I call it deranged for reasons that will become clear.

There are n! permutations of n objects, as we know. Of these re-arrangements a total D(n) are such that *no object returns to its original place*. Such a permutation is called a *derangement*. What is the probability of this event?

For example, if n people check their coats and these are randomly shuffled, what is the probability that no person retrieves his own coat when checking out? Or, again, if n distinct cards, arranged in some order, are thoroughly shuffled, what is the probability no card returns to the position it previously held?

To answer questions like this we require a formula for D(n) since the probability of a derangement is then simply D(n)/n! (The sample space consists of n! equally likely possibilities).

The following argument is due to Euler from 1751 based on the original formulation by Montmort in 1710 and the earlier solution by Johann Bernoulli (see Takacs [56] for the story of this problem).

The set of n objects are ordered as a, b, c, . . . The sample space is all n! possible permutations of these objects and our assumption is that all these orderings are equally likely. Note that D(1) is evidently 0, D(2) is 1 since the only two possibilities are ab and ba, and D(3) = 2 because the only derangements of abc are bca and cab.

In a derangement there are only n – 1 possibilities for the first place (it cannot be the object labeled as (a). Assume, without loss of generality, that the first letter is (b). There are then two cases to consider. Either the sequence begins with ba or not. In the first instance, the remaining n – 2 letters can be deranged in D(n – 2) ways (assume n > 2) and, in the alternate case, the remaining n – 1 letters (that now include the letter a among them) can be deranged in n – 1 ways. Altogether, then, there are

$$D(n) = (n-1)[D(n-1) + D(n-2)]$$

possibilities. Now rewrite this identity as

$$D(n)–nD(n-1) = –[D(n-1)–(n-1)D(n-2)] \qquad (15.3)$$

Except for the minus sign, each side of (15.3) is of the same form, one integer apart. To exploit this recursive property note that

$$D(n-1)–(n-1)D(n-2) = –[D(n–2)–(n-2)D(n-3)]$$

and therefore

$$D(n) - nD(n-1) = (–1)^2[D(n-2)–(n-2)D(n–3)].$$

Continuing in this manner eventually leads to

$$D(n) - nD(n-1) = (–1)^{n-2}[D(2)–2D(1)] = (–1)^n$$

It follows that

$$D(n) = nD(n-1) + (–1)^n \qquad (15.4)$$

I now use induction to establish that

$$D(n) = n![1–1/1! + 1/2! –1/3! + 1/4!– \ldots (–1)^n/n!] = n!\sum_{k=0}^{n}(–1)^k/k! \qquad (15.5)$$

This relation is certainly true for n = 1 and we assume it to be true for n = k. Then, using (15.4),

$$D(k+1) = (k+1)D(k) + (–1)^{k+1} = (k+1)[k!(1–1/1! + 1/2! - 1/3!$$
$$+ \ldots + (-1)k/k!] + (-1)^{k+1}$$
$$= (k+1)!\left[1 - 1/1! + 1/2! - 1/3! + \ldots + (-1)^k k! + (-1)^{k+1}/(k+1)!\right]$$

and this completes the induction.

Here is a surprising and counter-intuitive revelation. As n increases the terms within brackets in (15.5) approaches a constant value quite rapidly. In fact one obtains in the limit the infinite series representation to 1/e.

The quantity D(n)/n! is already quite close to e^{-1} when n = 20 and so we have the startling result that the probability of a derangement is virtually the same whether n equals 20 or n equals two million!

Instead of asking for the probability of a derangement, it is reasonable to ask for the probability of a coincidence. Let Z_n be the number of coincidences that occur in a random shuffle of n objects. Consider the event $Z_n = k$. Then pr $(Z_n = k)$ is obtained by choosing exactly k out of n objects to match (this can be done in C(n, k) ways) with the remaining n – k objects having no coincidence (recall that a derangement of n – k objects can be done in D(n – k) ways). Both of these events are independent and, since the sample space consists of n! possible shuffles, it follows that

$$\mathrm{pr}\ (Z_n = k) = C(n,\ k)D(n{-}k)/n! \tag{15.6}$$

Using (15.5) with n replaced by n – k we get

$$\mathrm{pr}\ (Z_n = \ k) = 1/k! \sum_{m\ =\ 0}^{n-k} (-1)^m/m! \tag{15.7}$$

In particular, when k = n one gets pr $(Z_n = n) = 1/n!$.

Another unanticipated result is that the average value of Z_n is exactly one, for any n. In fact $E(Z_n) = \sum k\ \mathrm{pr}(Z_n = k)$ summed from 1 to n and this equals $\sum k\ \ C(n,\ k)D(n - k)/n!$ which, in turn, equals $\sum[(n - 1)\ !\ /(k - 1)\ !\ (n - k)!]D(n - k)/(n - 1)!$ Now let k = m + 1 to obtain $\sum[(n - 1)\ !\ /m\ !\ (n - 1 - m)!]D(n - 1 - m)/(n - 1)\ ! = \sum C(n - 1,\ m)D(n - 1 - m)/(n - 1)!$ with the sum now being over m from 0 to n – 1.

We recognize this sum to be the same as $\sum \mathrm{pr}\ (Z_{n\ -\ 1} = m)$ and since the random variable $Z_{n\ -\ 1}$ ranges from 0 to n – 1 the sum must equal unity.

One can approximate $\mathrm{pr}(Z_n = k)$ by the Poisson distribution as we did for the binomial. The probability of a coincidence at some position for two random permutations is simply p = 1/n. Therefore the average number of coincidences in the Poisson approximation is $\lambda = np = 1$ which confirms what we have already shown. Moreover the probability of k successes (coincidences) is, approximately, λ^k exp (–1)/k ! = exp (–1)/k! and this equals (15.7). It now follows that the probability of a derangement (zero coincidences) is approximately exp(– 1) which confirms what was found earlier.

A MATLAB code to compute the probability of matching k items out of n, *Montfort.m*, is given in the Appendix I to close this section I want to discuss an example from a lottery as discussed in the article by Handley [66]. The same four digit lottery number 8092 came up on the same night in separate draws in the states of Massachusetts and New Hampshire and a gambling official reported that "there is a 1-in-10,000 chance of any four-digit numbers being drawn at any given time but the odds of it happening in two states at the same time are just fantastic." Evidently what was calculated was the probability of two independent samples, each having probability to be 10^{-4}, being multiplied by itself (one chance in100 million). But that would be incorrect. The fallacy here, as in the birthday problem discussed earlier, is that the officials focused on the coincidence of the specific number 8092 instead of matching *any* two four-digit numbers, the probability of which is just 10^{-4} or one chance in 10,000. Moreover they didn't distinguish between the occurrence of this event and it being noticed. Since the two states are neighboring both lotteries are reported in local papers but if the second state had been out west it is less likely the "fantastic" coincidence wild have been noticed.

An amusing take on coincidences is an opinion piece by TV personality Dick Cavett called, appropriately, *What Are The Odds of That?* [40].

Chapter 16
Turing's Evidence

Most accounts of Alan Turing's time at Bletchley Park focus on his achievements in breaking the Enigma code but until 2012 little was known about his use of Bayesian methods to solve problems in cryptology. I give here a brief summary of how Turing approached one such problem following the historical summaries in Zabell [39] and Simpson [38].

Although the actual context of the problem I'll discuss was somewhat different at Bletchley because it involved the configuration of an Enigma machine the essential idea can be conveyed by considering a classical coding scheme known as the Vigonere cipher that I'll describe next.

Write the 26 letters of the alphabet along the top and side of a 26 × 26 square for which row k represents a shift of the alphabet k places ton the right, modulo 26.

	a	b	c	d	e	f	g	...
a	B	C	D	E	F	G	H	...
b	C	D	E	F	G	H	I	...
c	D	E	F	G	H	I	J	...
d	E	F	G	H	I	J	K	...
e	F	G	H	I	J	K	L	...
f	...							
...	...							

Thus, row d is an alphabet shift of 4. To make use of this table a cryptologist will choose a keyword, say ABLE, which is repeated along the top row of a plaintext, say *the enemy is coming*, then encrypt the message by taking the first letter t and matching it to the first letter of the keyword to get the substitution t → A meaning that t is subjected to a 1 shift (row 1), namely t → U. Next plaintext letter is h which matches B in the codeword and this means the second row, namely a 2 shift from h, which is J. The e corresponds to L which initiates an 11-shift so that e → P. And so on:

E. Beltrami, *Digressions in Elementary Probability*, Quantitative Methods in the Humanities and Social Sciences, https://doi.org/10.1007/978-3-031-70051-4_16

Codeword	A B L E A B L E A B L E A . . .
Plaintext	t h e e n e m y is c o m i n g
Ciphertext	U J P . . .

The frequency of occurrence of the letters in a text written in English differ from letter to letter, such as

a	b	c	d	e	. . .
.082	.015	.028	.043	.127	. . .

If two different independent strings of plaintext are superimposed the probability that two letters coincide has a probability of p_k^2, where p_k is the frequency (probability) of the k*th* letter. Summing over all 26 disjoint events the probability of a match between any two letters is then $\sum pk^2 = .0667$ or about 6.7 coincidences per 100 letter pairs.

I now follow the treatment in Kahn [37], 377–380. We are given two or more ciphers which are superimposed assuming that each is generated from the same key word and that each string is lined up so that the two occurrences of that key are in sync with each other. In that case there also about 6.7 coincidences per 100 letters. This is due to the fact that since the superposition of texts are in phase two letters of each vertical pair will result from the same keyword substitution. Thus, whatever the plaintext, these letters will be identically enciphered. Thus, two c's may be coded as V at one point and Q at another, or a coincidence of a's becomes a coincidence of T's here and a coincidence of F's there. The total number of coincidences in the ciphertext remains the same as the number in the plaintext. That is, the coincidence will occur whether the letters are clothed in ciphertext disguises or not. The calculation does not ask for their identities. It merely notes their coincidence. The way Simpson puts it, "If two letters, one above the other in the enciphered messages, were the same – that is, iff they matched – and the alignment was correct then, since the keyword is the same, the corresponding letters in the original plain texts must also match."

On the other hand, "if the cryptograms are not properly superimposed so that the keys are not in step, any coincidences will result from different key letters operating on different plaintext letters to accidentally produce the same ciphertext letter. The coincidence will be caused by chance and we therefore expect, in this case, about 3.78 coincidences per 100 vertical pairs.

The reason is that each cipher letter has a probability of 1/26 of appearing since they are essentially uniformly distributed at random. If two such texts are superimposed then the probability of a match between any two is, due to each letter being chosen independently, 1/26 times 1/26 and so the probability of a match between the two superimposed ciphertexts is the sum of the probabilities of 26 disjoint events, namely $\sum(1/26)^2$ and this equals $1/26 = .0385$. An attack on this problem is to shift the ciphertext one place to the right (or left) at a time until one gets a maximum number of coincidences. As Kahn says "It is like shifting, an inch at a time, two identical picket fences with very wide pales and very narrow slits at irregular intervals. From time to time light will shine through when the slits coincide by chance. When the fences are correctly juxtaposed there is a burst of radiance and, similarly, the coincidences in plaintext will stand out when the cryptograms are correctly aligned. When the coincidences reach their highest level and begin to decline provides a clue to the length of the keyword, especially if this pattern repeats on a fairly regular basis."

Turing employed it in the following form. He wants to find the odds that two ciphertexts, encoded by the same keyword, are in unison. We know that if two message streams are correctly superimposed there will be a higher number of repeats then if they did not correctly fit. If the streams are randomly generated the probability of a coincidence is $1/26 = .035$, as was seen above. On the other hand, if the 26 letters are independently generated from a plaintext with letter k occurring with frequency p_k then $\beta = \sum pk^2$ summed from 1 to 26, has the value .067 (for English) and this is the probability two letters match.

Suppose the data D is that we obtain n repeats in a match of two ciphertexts of length N that overlap. The overlap can be viewed as a binary string of length N with 1 denoting a success, namely a match, and 0 a mismatch. We then have N Bernoulli trials having n successes. The probability of D given that we have a good fit (perfect alignment) is then $C(N, n)\, \beta^n(1 - \beta)^{N-n}$ and for a poor fit (random alignment) it is $C(N, n)\, (1/26)^n(25/26)^{N-n}$.

Therefore, in the odds form of Bayes Theorem, the likelihood or Baye's Factor is $Pr(D|\text{ good fit})/ Pr(D|\text{ random fit}) = (26\beta)^n[26(1 - \beta)/25]^{N-m}$.

Since $26\beta > 1$ and $(1 - \beta) < 1$ the Baye's Factor is an increasing function of n. Therefore, for some moderate number n of matches or repeats this factor exceeds one. Turing called this "the factor in favor of a hypothesis H provided by the evidence E."

Turing used Bayesian methods at a time (circa 1940) when they were little used among statisticians generally. He understood the importance of priors stating "The evidence concerning the possibility of an event occurring usually divides into a part about which statistics are available and a less definite part about which one can only use one's judgment". Priors that were significant in Bletchley's code-breaking efforts included partial information on the type of message, its possible content, and who was sending it. Interestingly, there was at Bletchley another group os statisticians who independently used Bayesian techniques including Edward Simpson who later became known for his counter-intuitive Simpson's Paradox.

Turing's factor can also be expressed as odds $(H \| E)$ / odds (H) and by takings logarithms we get

$$\log BF = \log \text{odds}\left(H \middle\| E\right) - \log \text{odds}\,(H). \tag{16.1}$$

which measures the increase in certainty based on new evidence. This is reminiscent of a different measure of certainty that is discussed in [3] based on the idea of entropy. I'll introduce it below.

To begin, a short digression on terminology. One *bit* of information is acquired by making a choice between two equally likely options, say 0 and 1. For a random binary string of length N there are 2^N equally likely options and a choice of one of them provides N bits of information given by $\log_2 2^N$. This is tantamount to saying that there are N bits of uncertainty. In general the binary digits occur with unequal probability and so one considers an average given by $H(1) = -\sum (n_j/N)\log_2(n_j/N)$ summed over $j = 1, 2$ where n_j is the frequency in which 1 or 0 appears in the binary string. This expression provides a measure of uncertainty or, to put it another way, the degree of randomness in terms of bits per symbol (the logarithm is base two whereas Turing utilized base ten).

By the Law of Large Numbers the fractions n_j/N approximate the probabilities p_j of a match or not depending on j and it is not difficult to show that H(1) is maximum, namely one, when $p_1 = p_2 = 1/2$. For unequal probabilities the binary string will exhibit redundancies that, by themselves, may not be a strong measure of non-randomness since repetitive blocks of some length k may appear indicating that the string has some degree of regularity. So now we look at diagrams 00, 01, 10, 11 within the string, blocks of length two, which appear with frequencies n_j/N and indices j running from 1 to 4 and we define H(2) as the sum over the four indices given by $-\sum (n_j/N) \log (n_j/N)$. Then $p_j = n_j/N$ approximates the probability of a particular diagram appearing. One can continue in this manner to obtain H (k) summed over 2^k k-grams and then compute a quantity

$$\mathrm{ApEn}(k) = H(k) - H(k-1) = -\sum p_j \log p_j + \sum p_j' \log p_j' \qquad (16.2)$$

with the first sum from 1 to 2^k and the second from 1 to 2^{k-1}. Note that there are $N + 1 - k$ possible k-grams in the string some of which may repeat and others not. The idea here is that we want to estimate the degree of randomness of a block of length k conditional on knowing its prefix of length $k - 1$.

This gives the *new information contributed by the last member of a block given that we know its predecessors within the block.* If the string is highly redundant, you expect that the knowledge of a given block already largely determines the succeeding symbol and so very little new information is accrued. At the other extreme, when the binary digits are randomly distributed, the difference ApEn will be small since each sum in (16.2) is close to unity. Therefore in this case the uncertainty, is maximal. Turing recognized that the presence of a fairly long k-gram, a run of k successive matches, is already an encouraging indicator that the string is not randomly generated. In fact we saw in Section One that long runs become increasingly unlikely in a random string.

If we let H be the hypothesis that the string of repeats is not random and E the evidence provided by an initial $k - 1$ segment of a k-gram then Turing's aphorism *the factor in favor of a hypothesis H provided by the evidence E* now applies equally well to ApEn (*Approximate Entropy*, the term used in [3]) and one notices that (16.2) bears a resemblance to (16.1). The difference is that the Bayesean factor increases with the presence of repeats while the ApEn factor, a measure of mutual information, decreases.

Chapter 17
The Paradox of Random Arrivals

This section and the one that follows are a bit more technical and makes use of continuous random variables unlike the rest of this book.

First a few preliminaries. A random variable X has an *exponential distribution* with parameter λ if its density function is $f(x) = \lambda \exp(-\lambda x)$ for $x \geq 0$ and zero otherwise. A simple integration by parts shows that E(X) = ∫ xf(x)dx integrated from 0 to ∞ equals 1/λ. A signature property of the exponential distribution that I will use later is that it is *memoryless*. This means that the likelihood of X achieving a time span of s + t units given it has already survived a time span of t units is equal to the probability of attaining at least another s time units. More abstractly we can write

$$P\left(X > s + t \middle\| X > t\right) = P(X > s) \text{forallnon} - \text{negatives}, \ t.$$

Using the definition of conditional probability this is the same as PX > s + t, X > t)/P (X > t) = P(X > s) or, equivalently, P(X > s + t||X > t) = P(X > t)P(X > s). Since we know that exp(−λ(t + s)) = exp (−λt) exp (−λs) this establishes the memoryless property for the exponential distribution.

For an exponential random variable X note that P(X ≤ t) = ∫ f(x)dx = λ ∫ exp (−λx)dx integrated from 0 to t. Using a straightforward integration this becomes 1 − exp (−λt) for t ≥ 0 and zero for t < 0, Conversely, by taking the derivative of P(X ≤ t) one immediately sees that f(x) = λ exp (−λx). This gives us two ways to characterize an exponential variable, a fact that will be used below.

The Poisson random variable was introduced informally in an earlier section. I now get to be a bit more formal and describe a counting process in which events occur sequentially with the number of events taking place by time t is denoted by N (t) for t ≥ 0. The probability that N(t) = k is assumed to be given by given by a Poisson density $p(t) = (\lambda t)^k \exp(-\lambda t)/k!$ Note, in particular, that P(N (t) = 0) = exp (−λt). Using this density function one can compute the expected value of N(t) in the same manner as we did for the discrete Poisson variable in the Tutorial and find that E(N(t)) = λt.

E. Beltrami, *Digressions in Elementary Probability*, Quantitative Methods in the Humanities and Social Sciences, https://doi.org/10.1007/978-3-031-70051-4_17

The basic assumptions of the Poisson counting process is that the number of events that occur in disjoint time intervals are *independent* and that the time gaps between events are *stationary* in the sense that they depend only on the length of the inter-event interval and not on when it occurs. That is, for any $s > 0$ and $t_1 < t_2$, the probability of $N(t_2 + s) - N(t_1 + s)$ is the same as the probability of $N(t_2) - N(t_1)$.

Now let T_n be a sequence of inter-event (inter-arrival) times, n = 1, 2, . . . Then $P(T_1 > t) = P(N(t) = 0) = \exp(-\lambda t)$ because $T_1 > t$ occurs if no Poisson events take place by time t. It follows, from what was said earlier, that T_1 has an exponential distribution with mean $1/\lambda$. We now observe, using the property of conditional expectation discussed in the Tutorial, that

$$P(T_2 > t) = E\left[P\left(T_2 > t \| T_1\right)\right].$$

But from independence and stationarity of event gaps one obtains $\text{prob}\left(T_2 > t \| T_1 = s\right) = P\left(\text{zero events in } (s, s+t) \| T_1 = s\right) = P(\text{zero events in}(s, s+t)) = \exp(-\lambda t)$. It follows that T_2 is also exponentially distributed with the same mean and it is independent of T_1. The same reasoning shows the T_n constitutes a sequence of independent event gaps each of which has the same mean $1/\lambda$.

I now come to the paradox, sometimes referred to as the *Inspector's Paradox.* My reasoning follows the clever argument in Ross [13].

Suppose someone arrives at some random time t to observe an ongoing Poisson process with its recurrent exponential gaps between events. An example would be some electronic device that operates continuously until it breaks down and is then immediately replaced to begin again. It could be something as simple as a failure of a light bulb or a battery. A randomly arriving observer, a machine inspector let us say, waits until the next failure. Or, for something different, a person arrives at a bus stop at some random moment to wait for the next bus in which the exponential inter-arrival gaps for each successive bus are also from a Poisson process. The remarkable thing is that the *average gap length of those gaps randomly penetrated by the observer are longer than the average* $1/\lambda$ *of the gaps in general.*

To see this note that the length of the penetrated gap is the sum of two parts: a length L_1 from the time of random arrival at time t until the next bus appears or the next event occurs and a length L_2 which is the interval between the previous arrival and t or, looking backward, the time from t to the previous arrival. By the memoryless property of the exponential and the independence of L_1 and L_2 and since the event process is Poisson we know that L_1 is an exponential random variable with mean $1/\lambda$ and, as discussed earlier, $P(L_1 \leq t) = 1 - \exp(-\lambda t)$.

As for L_2, let x be any time before t. To claim $P(L_2 > x)$ means that no arrivals have occurred between x and t, namely no arrivals within the interval $[t - x, t]$. Therefore $P(L_2 > x) = P(0 \text{ events in } [t - x, t])$ for $x \leq t$, with a probability of zero if $x > t$. The last sentence implies, given that we have a Poisson arrival process, $P(L_2 > x) = \exp(-\lambda x)$ for $x \leq t$ and equals zero if $x > t$.

The previous sentence can be restated to say $P(L_2 \leq x) = 1 - \exp(-\lambda x)$ for $x \leq t$ and equals one if $x > t$. So, if the penetration time t is large enough L_2 is an approximately exponential random variable with mean $1/\lambda$.

The average of $L_1 + L_2$ is therefore the sum of the individual averages which is approximately **$2/\lambda$** or *twice the usual gap length!* A heuristic explanation for this apparent paradox is that *a randomly arriving observer is more likely to penetrate a longer than average gap interval than a shorter one.*

In the Appendix I provide a MATLAB code *Poisson's* that generates n exponential inter-arrival times from a Poisson process with mean λ as well as the cumulative sum of these arrival times. Since $P(T > t) = \exp(-\lambda t)$ is a probability between zero and one it suffices to choose a random number v between these limits whose values are uniformly distributed. These indicate the possible values of an exponential arrival and if we solve for t by taking a logarithm one obtains $t = -\log v/\lambda$. This is the expression used in the code. For example, if n is 3 and $\lambda = 1.5$ then we obtain .2427, .7657, and .0340 as three such arrival times.

Chapter 18
The Inverse Square Root Law

A spatial Poisson process is defined in a way similar to the manner by which a temporal process was treated in the preceding vignette. Suppose events take place in the plane at random and that if S is any subset of the plane, then N(S) counts the number of events that occurs within S. If S and S′ are disjoint subsets then we require that N(S) and N(S′) be independent random variables (*independent increments*) and that N(S) depend only on the area of S and not on its position or shape (*stationary increments*).The spatial counting process is called Poisson if

$$\mathrm{prob}(N(S) = k) = (\gamma A(S))^k \exp(-\gamma A(S))/k!$$

where $k = 0, 1, \ldots$ and A(S) is the area of S. The rate constant γ is easily shown to be the average number of events per unit area, using an argument analogous to that employed in the temporal case where the average number of events per unit time was designated by λ.

A large region S of area A(S) has N emergency response vehicles (such as fire trucks, ambulances, or police patrol cars) scattered at random according to a spatial Poisson distribution at rate γ. The rate constant is the average number of vehicles per unit area and is estimated to be N/A(S). Alarms arrive as a Poisson process at rate λ and any response unit that is dispatched remains busy for a length of time that is exponentially distributed with a mean duration of 1/μ. We assume that calls arrive over time during peak alarm periods during which the mean λ can be considered to be constant. The more quiescent periods of the day, if there are any, are less interesting since the demand for service can be satisfied more readily.

An incident occurs somewhere at random, uniformly distributed within S, and the closest available unit is dispatched. Implicit here is that alarms are just as likely to happen in one part of the region as another. This implies that S is fairly homogeneous in terms of hazards, as would be the case in a number of urban and even suburban areas. For the moment it is also assumed that all N units are available, a restriction

E. Beltrami, *Digressions in Elementary Probability*, Quantitative Methods in the Humanities and Social Sciences, https://doi.org/10.1007/978-3-031-70051-4_18

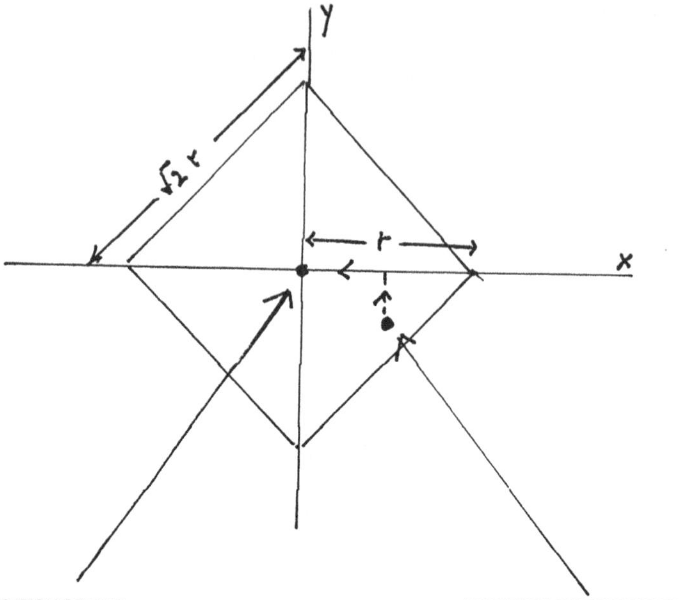

Fig. 18.1 Distance using the right-angle metric

that will be relaxed later to account for the fact that some responders may be busy at other alarms.

Let us think of S as a portion of a metropolis in which the street network is a grid of criss-crossing intersections. This means that travel to an incident is not along the shortest path between two points, or "as the crow flies," but along a less direct route. We want to make this idea more precise by introducing a measure of distance r between the origin and a point having coordinates (x, y) to be $|x| + |y|$. Travel in terms of this metric is along a horizontal distance |x| followed by a vertical distance $|y|$, which is different from the conventional Euclidean metric distance defined by $\sqrt{(x^2 + y^2)}$. In an actual street pattern, the distance traveled would be somewhere between these extremes but at the risk of being too conservative we adopt the first of these distance measures, namely $|x| + |y|$.

The locus of points in the x, y plane that are within a distance r of the origin is of course a circle of radius r, when using the Euclidean metric, and the area is πr^2. But, as a little thought will show, the corresponding locus of points that are within a distance r using $|x| + |y|$ is now a square tilted by 90 degrees with sides of length $\sqrt{2}r$ and area $2r^2$ as illustrated in the figure below (Fig. 18.1).

If the travel speed of a vehicle is roughly constant, then response time is proportional to travel distance and so from now on we work with distance traveled rather than elapsed time.

At this juncture, we are able to compute the probability distribution of the travel distance D between an incident and the closest responder. Choose the coordinate system so that the incident is located at the origin. Since the responding vehicles are Poisson distributed in the region, the probability of there being none within a distance r of the incident is simply $\exp(-2\gamma r^2)$, using $|x| + |y|$, given the fact that area is $2r^2$. Therefore the probability that at least the closest unit is within a distance r is given by

$$F(r) = \text{prob}(D_1 \leq r) = 1 - \text{prob}(D_1 > r) = 1 - \exp(-2\gamma r^2)$$

The density function of the random variable D is obtained from by differentiating F (*r*) and is given by $dF(r)/dr = 4\gamma r\exp(-2\gamma r^2)$. The expected value of D is now computed by the integral from 0 to ∞ given by

$$E(D) = \int rF'(r)dr = 4\gamma \int r^2 \exp(-2\gamma r^2)dr$$

which is evaluated by rewriting this as:

$$E(D) = -2\gamma\, d/d\lambda \int \exp(-2\gamma r^2)dr$$

Now make the substitution $x = \surd(2\gamma)r$ to obtain

$$\int \exp(-2\gamma r^2)dr = 1/\surd 2\gamma \int \exp(-x^2)dx$$

and from a table of integrals we find that the last integral has the value $\surd\pi/r$. Putting all this together gives

$$E(D) = -\gamma\surd\pi/2\ \ d/d\gamma(1/\surd\gamma) = 0.627\ \surd 1/\gamma = c/\surd N. \qquad (18.1)$$

For som suitable constant c.

We therefore see that the expected travel distance is inversely proportional to the square root of the density of available units. This relation persists under a variety of assumptions about the distribution of response vehicles in the region and the metric one employs. It even continues to be valid if the *k*th closest unit is dispatched instead of the nearest one although I don't show the somewhat messy computation here.

Empirical verification of the inverse square root relation has been obtained by plotting actual response time data against the number of available units (see the paper by Kolesar and Blum [43] which is confined to Fire Department operations in New York City), and it is remarkable that the relation persists in spite of the several tenuous assumptions that were made in its derivation, some acceptable, others perhaps less so, such as constant vehicle speed, uniform alarms in space, exponential service times, and a rectangular street grid.

The robustness of the relation (18.1) between response time and the number of available units under a variety of conditions compels us to call it the *inverse square root law*.

I want to comment on how the square root law changes when one recognizes that not all N units are available to respond to an emergency because some of them may be busy on other calls. Let $q = N - m$ where m is a random variable that denotes the number of busy and therefore unavailable responding units. It depends on call and vehicle service rates and an exact value for its expected value can be obtained using ideas from queuing theory. This won't be necessary here since I simply want to see the effect of q on (18.1). Since q is a random variable

$$E(D_1) = E\big(E(D_1 \parallel q)\big) \text{ where } E(D_1 \parallel q) = c_1/\sqrt{q}.$$

The functions $1/\sqrt{q}$ and $\sqrt{q}$ are negatively correlated and so covariance $C(\sqrt{q}, 1/\sqrt{q}) = E(1) - E(\sqrt{q})E(1/\sqrt{q}) < 0$ from which $E(1/\sqrt{q}) > 1/E(\sqrt{q})$. Moreover the variance of $\sqrt{q}$ is $E(q) - E^2(\sqrt{q}) > 0$ and this shows that $E(\sqrt{q}) < \sqrt{E(q)}$ and consequently the expected value of the function $c/\sqrt{q}$, namely E(D), is greater than

$$c/\sqrt{E(q)} = c/\sqrt{(N - E(m))}.$$

If one accepts this lower bound as the worst case scenario then we again obtain an inverse square root law with N replaced by the average number of available units, as one might have surmised.

Applications by the Rand Group to New York City Fire Department operations were discussed in a NY Times article [71] and in the book [76]. From this last work one gets yet another application of the Poisson distribution. This example may be of interest since it involves temporal rather than spatial occurrences. It originates from a RAND study of Fire Department operations in New York City around 1970. Fire alarms typically vary by time of day and by season. The author's of the study chose five consecutive Friday summer evenings in a region of the Bronx between 8 and 9 pm during which there were $n = 55$ alarms. The five hour period was divided into 15 min segments for a total of $M = 20$ intervals so that λ can be estimated as $n/M = 2.75$, the number of alarms per unit time. To check whether the data is consistent with a Poisson distribution the following Table gives the number of intervals N_k having exactly k alarms followed by the quantities $N\,P_k$:

k	0–1	2	3	4 or greater
M_k	5	5	4	6
M *times* P_k	4.80	4.83	4.43	6.94

We see another close fit suggesting that the fire alarms arrived at random times. This was useful to the investigators since they were attempting to study delays in Fire Department response times and this depends on having Poisson arrivals of alarms.

Chapter 19
Runs, Again

> We shall encounter theoretical conclusions which are not only unexpected but actually come as a shock to intuition and common sense. (Feller [4])

The opening section of these notes was an attempt to clarify our intuition about sequences of Bernoulli trials and I now close by augmenting that discussion with a different example of a sequence of Bernoulli events by looking at a simple coin-tossing game between two opponents. The first person wins a dollar from the other player if the binary digit is one (indicating a head); otherwise he losses a dollar to his opponent when the digit is zero (tails). Using a fair coin one expects, as a result of the Law of Large Numbers, that the average gain (or loss) after a large number of tosses to tend to zero as n increases.

However, there is a surprising twist to this game if one looks at the total gain (or loss) of each player, and not at the average. It turns out that one or the other player will be in the lead most of the time! This curious and counterintuitive phenomenon can be verified by simulating such a game which records the *fluctuations of total gain or loss*. A typical game of 10,000 tosses is illustrated in the figure below (plotted as every tenth toss) and it turns out that *one of the players is in the lead 7416 times, roughly 75% of the tosses*!. The average gain, however, is only .01246, close to zero. I simulated this game using a MATLAB code *Cointoss.m* that is provided in the Appendix.

A casual observer might understandably conclude that one of the players is cheating but, as will be seen, such asymmetry between gain and loss is very much in the nature of randomness. As [Feller [4], 142] puts it, "It is usual to read into the Law of Large Numbers things which it definitely does not imply. If Peter and Paul toss a perfect coin 10,000 times it is customary to expect that Peter will be in the lead roughly half the time. *This is not true*....there does not exist any tendency for the periods of lead to equalize. The Law of Large Numbers asserts only that in a large number of *different* coin-tossing games the frequency of those in which heads lead

E. Beltrami, *Digressions in Elementary Probability*, Quantitative Methods in the Humanities and Social Sciences, https://doi.org/10.1007/978-3-031-70051-4_19

is, at any given moment, close to ½. Nothing is said about fluctuations of the lead within a *fixed* game."

Let the total winnings after n tosses to be the sum $S_n = \pm \sum 1$, taken from 1 to n, where plus indicates a favorable winning head and minus a losing tail. The values of S_n generate a random walk to the right and left of zero that can be represented above and below a time axis that is marked by successive flips of the coin, as indicated in the previous fig. I am interested in the first return to the origin, namely those times k for which $S_k = 0$ and $S_r < 0$, for $r < k$. This is where the players break even. It is evident that S_k can be zero only when k is even, namely $k = 2$ m, for some integer m. Note that a return to the zero axis does not necessarily entail a crossing of the axis since the path can ricochet back to the same side it came from.

As an interesting aside I compute the probability u_{2m} of a return to the origin at time 2 m (not necessarily the first such return) note first that there are exactly 2^{2m} possible paths of length 2 m that begin at the origin (0, 0) and, of these, there are C (2 m, m) paths having an equal number of favorable and unfavorable flips as would be necessary in order to have $S_{2m} = 0$. It follows that $u_{2m} = C(2m, m)/2^{2m}$. The value of u_0 is taken to be unity.

I quote again from Feller: "This topic is pursued by investigating how often the lead is likely to change for an arbitrary path and we reach conclusions that play havoc with or intuition. It is generally accepted that in a prolonged sequence of tosses Peter should lead about half the time and Paul the other half. This is entirely wrong, however. In 20,000 tosses it is about 88 times more probable that Peter leads in all 20,000 games than that each player leads in 10,000 trials. In general the lead changes at such infrequent intervals that intuition is defied. No matter how long the series of tosses the most probable number of changes in lead is zero. If many coins are tossed n times each, a surprisingly large proportion of them will leave one player in the lead almost all the time; and in very few cases will the lead change sides and fluctuate in such a manner that is generally expected of a well behaved coin."

To explain Feller's comment consider a succession of n separate games between Peter and Paul. Let A be the event "Paul is in the lead at the m *th* toss, that is Paul wins in a game of exactly m tosses for any m (Feller's "at any given moment"). Thus A is equivalent to $S_m > 0$ in terms of the total winnings in m trials. Consider Bernoulli trials in which $X_k = 1$ if event A takes place in game k, and 0 otherwise. Pr $(X_k = 1) = ½$ since Peter and Paul are symmetric. The Law of Large Numbers tells us that the average $1/n \sum X_k$, over all n games, is roughly ½ for large n. Thus the fraction of games in which Paul leads at the m th toss, namely the percentage of games in which we find that Paul is winning at the m th step, is about ½. The same is true for Peter and this is what our intuition erroneously leads us to believe about fluctuations within a single game'.

I close by mentioning that the winnings of Peter and Paul even out an infinite number of times, namely the sums S_n have *recurrent zeros* in the sense that the probability S_n being zero infinitely often equals 1. A proof would take us beyond elementary probability (Fig. 19.1).

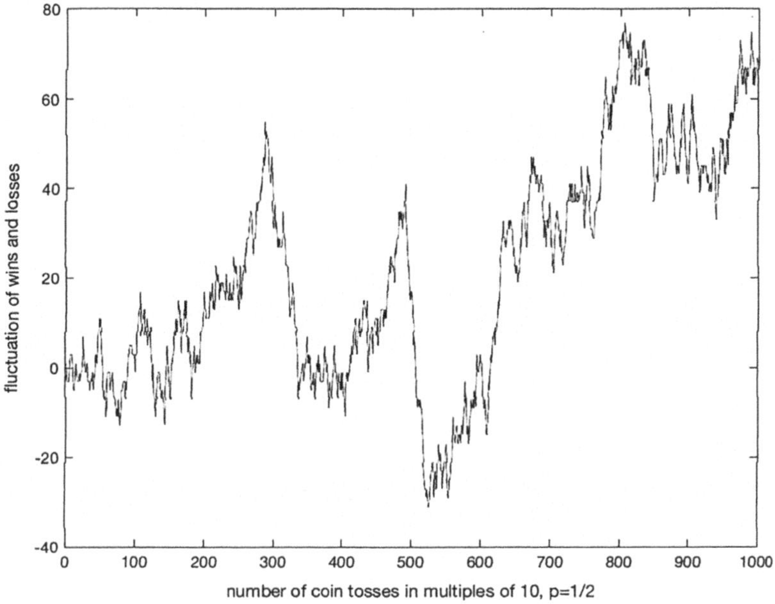

Fig. 19.1 Coin tosses

Chapter 20
Losing Strategies That Win

I want to describe another coin-tossing game that, at first glance, one would expect to result in a losing strategy. Quite surprisingly, quite the opposite is true, as discovered by physicist Juan Parrondo in 1997 and reported in a New York Times article [64] just a few years later.

The discussion below is based on a paper by Parrondo and Dinis [62]. I begin with a brief review of the paradox. There are two distinct coin-tossing games, call them A and B. The first is quite simple and returns a dollar gain if the toss results in a head, with probability p_A, and the loss of a dollar if the coin is a tail. Since $p_A < ½$ the coin is biased In favor of a loss and so this is evidently a losing game; the expected gain decreases with each toss and in the long run any initial capital will be lost.

Game B is a bit more complicated. Here we have two biased coins. Let x(n) be the accumulated capital at the nth toss for n = 1, 2, ... If x(n) is a multiple of 3 then we toss a coin whose probability of a head is $p_- < 1/2$. We call this the "the "bad" coin. Otherwise, when x(n) is not a multiple of 3, we toss a "good" coin whose probability of a head is $p_+ > 1/2$. It turns out, remarkably, that game B is also losing as will be established. Finally, I'm able to get to Parrondo's paradox by combining games A and B. It works like this. Toss a fair coin. If it comes up head play game A; otherwise play game B so, in effect, one randomly switches between games. We now have, to our wonder, a winning game obtained from two losing games! This is the paradox.

To understand how all this comes about and, in particular, the nuances of the B game I need to introduce the concept of *Markov Chains,* named after the Russian mathematician A. Markov.

Consider a random process over a sequence of steps indexed by n = 0, 1, 2, ... which can find itself in one of N integer states. For each n there is a random variable that takes on one of these N values with the probability of being in state j at the nth step being conditional on residing in state i in the previous step. That is, a move to state j is dependent on the current state although not on the previous history of the process. Moreover this probability is independent of the step we happen to stumble on when the process is observed. So we can define, for all n and any i and j, p_{ij} = P

E. Beltrami, *Digressions in Elementary Probability*, Quantitative Methods in the Humanities and Social Sciences, https://doi.org/10.1007/978-3-031-70051-4_20

(being in state j at n*th* step given that process is in state i at previous step). Clearly $p_{ij} \geq 0$ and $\sum p_{ij} = 1$ when summed over j since a move from a given state i is always to some other state j.

I trust the reader recalls the notion of a *vector* of N numbers (represented as a row or column) and that of an N by N *matrix* of N^2 numbers.This allows us to define a state vector x(n) whose j*th* component is the probability of being in state j at instant n and also a transition matrix T whose entry in row i and column j is p_{ij}. The row sums are evidently equal to one and the matrix product of x(n) and T results in x(n + 1). For example, if x(n) = (0.1/2, 1/2) and if the three rows of T are (0, 1/3, 2/3), (2/3, 0, 1/3), (0, 1, 0) then by the *Law of Total Probability* (as discussed in the Tutorial) the first component of x(n + 1) becomes (0 × 0) + (2/3 × 1/2) + (0 × 1/2) = 1/3 and similarly for the other components.

I assume, in what follows, that every state can be reached from every other state. Then a basic theorem of Markov chains asserts that as n increases x(n) tends to a vector π^* having N constant non-negative components summing to one. These components represent the limiting probabilities of being in each of the N states for large n. Moreover $\pi^* = T\pi^*$ or equivalently $(T-I)\pi^* = 1$, where I denotes the N by N identity matrix (ones along the diagonal and zeros everywhere else).

A proof of this theorem requires some knowledge of elementary real function theory and some linear algebra and so I omit it here but an ambitious reader can consult a short and sleek proof by Maltese that is available in [63].

Let's take up the story of the coin-tossing games A and B in more detail. Vector x (n) denotes the total earnings at the nth toss. This is a random variable having integer values. The outcome in game B is now dependent on the result of the previous toss since it can lead to x(n) being a multiple of 3. This is not the case in game A. Pictorially the games are (using specific but somewhat arbitrary choices for the probabilities) illustrated below where p_- denotes the probability of winning with the "bad" coin and p_+ for winning probability with the "good: coin:

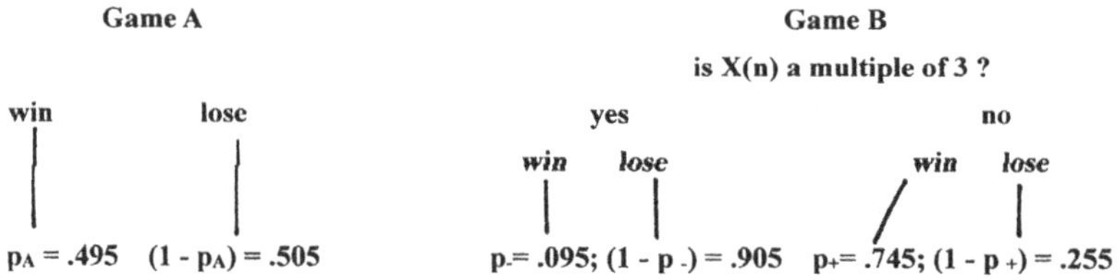

It will be convenient to reduce the total winnings to just one of three possible states by defining y(n) = x(n) *modulo* 3. Modulo is a mathematical term employed for what is essentially clock arithmetic. In telling time, for instance, the fourteenth hour is treated as if it is two o'clock and eighteen o'clock as if it were six. In fact all hours that differ by a multiple of 12 are considered to be the same. In an analogous manner the accumulation of earnings is either 0, a multiple of 3 modulo 3, or 1 (a multiple of 3, modulo 3, plus 1) or 2 (a multiple of 3, modulo 3, plus 2).

The three categories correspond to 0, 3, 6, 9, . . . or 1, 4, 7, 10,. ..or 2, 5, 8, 11,. . . .The vector y(n + 1) clearly depends on y(n) and defines a Markov Chain. The corresponding probabilities of being in each of the three possible states 0, 1, or 2 at each toss are the components of the probability vector y(n).

I need now to write down the transition matrix giving the probabilities of moving between states in game B from one toss to another. Let π_0*, π_1*, π_2* be there limiting probabilities for y(n) to be found in each of the three states (modulo three). The probability of winning game B is then

$$p_{win} = \pi_0 * p_- + (1 - \pi_0*)p_+ \tag{20.1}$$

The goal now is to find π^* in order to compute, using (20.1), the likelihood game B is a winning proposition or not.

There are two possibilities for y(n) = 2 to occur depending on whether the capital x(n) is a multiple of three or not. If y(n − 1) = 0 we lose at the next toss with probability $1 - p_-$ or else y(n − 1) = 1 and we win with probability p_+ at the next toss. Writing the k*th* component of y(n) as y(k, n) we obtain

$$y(2, n) = (1 - p_-)y(0,\ n - 1) + y(1,\ n - 1)p_+$$

Next, there are two possibilities for y(1) to happen and, reasoning as we just did, if y (n - 1) = 0 then the next move is to win with probability p_- or to be in state 2 and lose with probability $1 - p_+$. Thus

$$y(1, n) = y(0, n - 1)p_- + y(2, n - 1)(1 - p_+)$$

Finally there are two possibilities to be in state 0 depending on whether one is in states 1 or 2 in the previous step. By an argument similar to the two previous cases we get

$$y(0, n) = y(1, n - 1)(1 - p_+) + y(2, n - 1)p_+$$

This leads to the system of equations expressed in matrix form as

$$y(n) = Ty(n - 1) \tag{20.2}$$

In which the transition matrix T is

$$T = \begin{pmatrix} 0 & 1 - p_+ & p_+ \\ p_- & 0 & 1 - p_+ \\ 1 - p_- & p_+ & 0 \end{pmatrix}$$

Since y(1) = Ty(0) then y(2) = Ty(1) = T^2y(0) and by iterating this recursion one finds that y(n) = T^ny(0) ior any initial state y(0). Since this defines a Markov chain it follows that y(n) tends to some vector $\boldsymbol{\pi}* = (\boldsymbol{\pi}_{0*}, \boldsymbol{\pi}_{1*}, \boldsymbol{\pi}_{2*})$ as n tends to infinity, independent of the initial vector. This expresses the long term probabilities of being is states 0, 1, or 2 (modulo 3) as the case may be, namely the fraction of time the random process, which is the value of the total accumulation of earnings, resides in each of these states. Moreover π^* satisfies the vector relation $\boldsymbol{\pi}* = T\boldsymbol{\pi}$, namely $(T - I)\boldsymbol{\pi}* = 0$.

To find π^* we impose the additional requirement that because we are dealing with mutually exclusive probabilities,

$$\pi_0 * + \pi_1 * + \pi_2 * = 1. \tag{20.3}$$

From (20.1) and (20.2) it is evident that we need to solve a system of four equations in three unknowns, namely $(T - I)\boldsymbol{\pi}* = 0$, together with (20.3):

$$\begin{pmatrix} -1 & 1-p_+ & p_+ \\ p_- & -1 & 1-p_+ \\ 1-p_- & p_+ & -1 \\ 1 & 1 & 1 \end{pmatrix} \begin{pmatrix} \pi_{0*} \\ \pi_{1*} \\ \pi_{2*} \end{pmatrix} = \begin{pmatrix} 0 \\ 0 \\ 0 \\ 1 \end{pmatrix}$$

The matrix above has maximum rank and we can obtain a unique solution (I did this by a simple MATLAB command) which, after substituting in the values of p_- and p_+ leads to

$$\pi * = (.3836, .13, .4621).$$

Returning to (20.1) it is now possible to compute p_{win} explicitly by using $\pi_0{}^* = .3836 : p_{win} = .3838 \bullet .095 + (1 - .3836) \bullet .745 = .4957 < 1/2$. Thus, B is a losing game!

Finally, I'm able to get to Parrondo's paradox by combining games A and B. It works like this. Toss a fair coin. If it comes up head play game A; otherwise play game B. The two games are now interdependent and by conditioning on whether A or B is played (each with probability 1/2) one obtains a new transition matrix, call it P. P is found by combining the matrices for T and the matrix T' that applies to game A. The latter is simply

$$T' = \begin{pmatrix} 0 & 1-p_A & p_A \\ p_A & 0 & 1-p_{A.} \\ 1-p_A & p_A & 0 \end{pmatrix}$$

We arrived at T' by considering that for an increase of total earnings to a multiple of three (and this includes zero) the previous state is either a multiple of one and we lose with probability $1 - p_A$ or it is a multiple of two and we win with probability of p_A. Very similar reasoning applies in how earnings increase to a multiple of one or two as the case might be and, because there are no "good" and "bad" coins to worry about in game A, it is straightforward to obtain the transition matrix T' given above.

Putting T and T′ together one obtains

$$P = 1/2\begin{pmatrix} 0 & (1-p_A)+(1-p_+) & p_A+p_+ \\ p_A+p_- & 0 & (1-p_A)+(1-p_+) \\ (1-p_A)+(1-p_-) & p_A+p_+ & 0 \end{pmatrix}$$

As before, we need to find a solution of a system of four equations in three unknowns to find $\boldsymbol{\pi}^*$for the combined games. Since $\boldsymbol{\pi}^*$ is a probability vector the sum of its three components is unity, as before. So now the equations are $(T - I)\pi * = 0$ and $\pi_0{}^* + \pi_1{}^* + \pi_2{}^* = 1$ and we find that the following matrix

$$1/2\begin{pmatrix} -2 & (1-p_A)+(1-p_+) & p_A+p_+ \\ p_A+p_- & -2 & (1-p_A)+(1-p_+) \\ (1-p_A)+(1-p_-) & p_A+p_+ & -2 \\ 2 & 2 & 2 \end{pmatrix}$$

multiplies $\begin{pmatrix} \pi_0^\star \\ \pi_1^\star \\ \pi_2^\star \end{pmatrix}$ to give $\begin{pmatrix} 0 \\ 0 \\ 0 \\ 1 \end{pmatrix}$

Solving this system of equations, again using a simple MATLAB command, a unique solution is found to be $\pi * = (.3451, .2541, .4008)$. The probability of winning in this random combination is now $p_{win} = \pi_0{}^* \bullet (p_- + p_A)/2 + (1 - \pi_0{}^*) \bullet (p_+ + p_A)/2 = .3451 \bullet (.590/2) + (1 - .3451) \bullet (1.24/2) = .5078$. But $.5078 > 1/2$ and *we have a winning game*!

The first figure below, a simulation, shows the result of playing game B for 200 tosses, starting with an initial capital of 100. The loss is apparent. The next figure exhibits the growth of capital in Parrondo's game, also a simulation, in this case 2000 repetitions of 100 tosses each. Now growth is evident. I wrote a MATLAB code *Parrondo.m* for these simulations and this can be found in the Appendix (Figs. 20.1 and 20.2).

Parrondo's Paradox is a metaphor for the idea that two negative results can combine to form a positive one. It is reminiscent of the statistical curiosity known s Simpson's Paradox that was discussed in an earlier vignette in which two negative outcomes can combine to generate a success. Other suggestive examples can be found, one of which is the intriguing possibility that by rebalancing a financial portfolio in a suitable manner two neutral or even losing investments can combine to provide financial growth. It is known as *Shannon's Demon* since it was first proposed by Claude Shannon, the mathematician responsible for information theory and a noted investor. I follow the account in [a book by Poundstone [42] where you

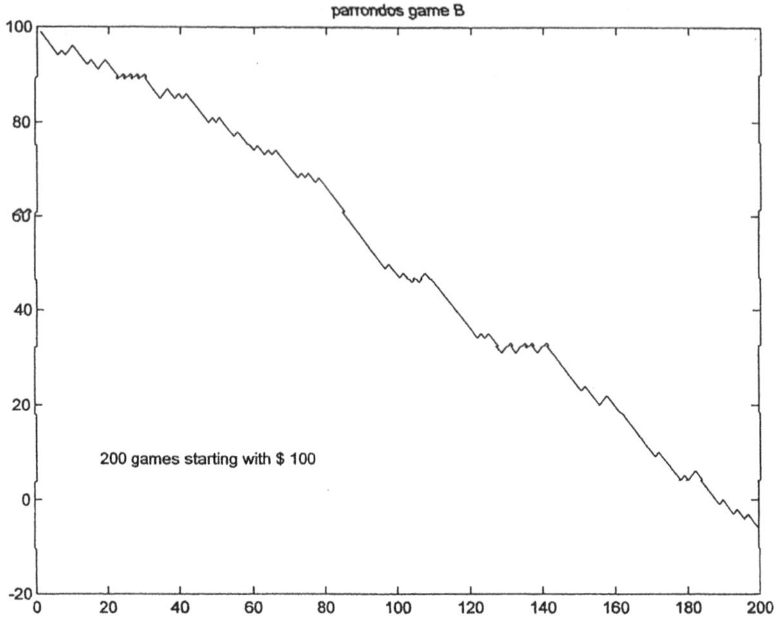

Fig. 20.1 Playing with 200 tosses

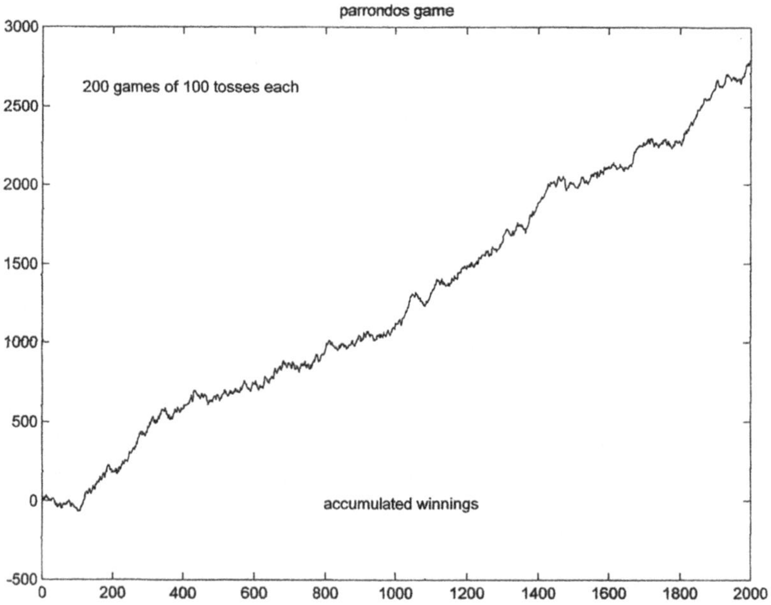

Fig. 20.2 Growth in capital in Parrondo's Game

also get an explanation for the provocative epithet "Demon ". His book was published before Parrondo's work became known but the connection to this paradox is acknowledged in [65].

A basic assumption about stocks is that their movement up or down is essentially a random walk with no preference for either direction. Suppose an investor has a thousand dollars, half in cash, half in stocks, and the strategy is to buy and hold. Over time the value of the portfolio will fluctuate up or down about a mean of one thousand with no growth. What Shannon suggested instead is that each day the portfolio is rebalanced so that the amount in cash and stocks remain equal. So, if on the first day the stocks lose half their value, let us say, one now has 500 in cash and 250 in stock.

The new investment strategy is to put 125 of the cash into stocks giving 375 equally in both assets. Then, next day if stock now doubles the portfolio consists of 750 in stock and 375 in cash for a total of 1125, a net gain. The same thing happens if the stock went up the first day and declined on the second. There is an even chance of up or down so over time the portfolio will jiggle up or down but on average will grow and not remain stagnant.

Moreover, the jiggling, namely the departure from the mean or as we know it from earlier vignettes, the variance or risk, in now less as can be seen in a computer simulation exhibited in the figure below titled *demon wealth* which shows the outcome of 200 trades in which the lower curve is the result of "buy and hold" and the upper is Shannon's rebalancing. The Shannon strategy is somewhat counter-intuitive because one puts cash into a losing stock and sells a winning stock.

Actually the mean of a group of stocks generally does show a trend, up or down, for extended periods but Shannon's point is that even if stock movement were completely random with a constant average over time it is still possible in principle, assuming modest trading fees, to accumulate wealth contrary to what many investors are prepared to believe.

The coded MATLAB simulation is available in the Appendix as *Shannon.m* (Fig. 20.3).

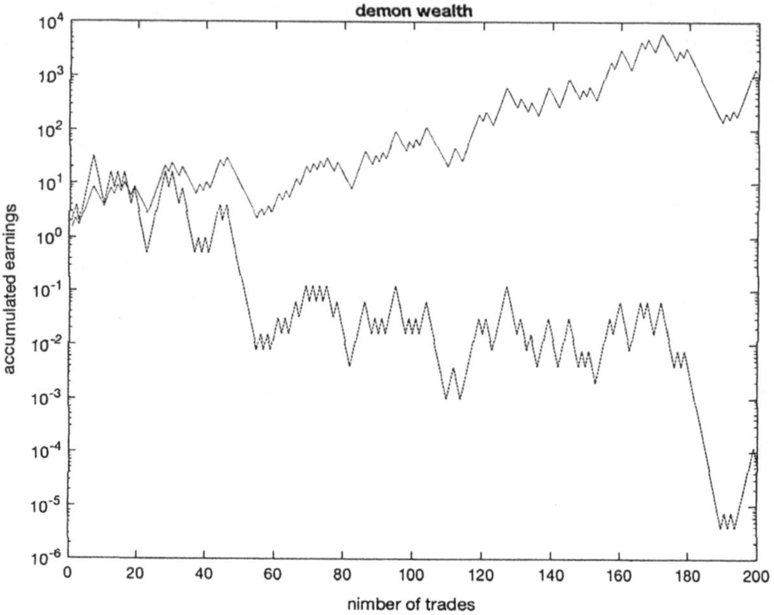

Fig. 20.3 Shannon strategy

A Tutorial on Elementary Probability

This is a brief introduction to some of the basic probability concepts that are used often in the main part of the book. It is intended to be mostly a review for readers who have had some prior exposure to the mathematics of probability and statistics and who welcome a refresher. I largely restrict myself to discrete random variables since that is all we need for most of the book. Slightly more advanced ideas are developed within the various vignettes as we go along.

Equation numbers begin with (1) and are unrelated to the main part of the book.

Counting Principles

We distinguish the act of *selecting*, also called a *combination*, without regard to order, k objects out of n, from that of *arranging*, namely a *permutation*, k objects out of n. The number of ways of accomplishing a combination we denote by C(n, k) and the number of ways for a permutation by P(n, k). C(n, k) is called a *binomial coefficient.*

For example, given 3 objects labeled {a, b, c} there are three ways of selecting two of these objects without regard to order, namely {a, b} or {a, c} or {b, c} so that C(3, 2) = 3. However, the number of arrangements of two of these objects out of the three given, namely P(3,2), is six since it consists of {ab}, {ba}, {ac}, {ca}, {bc}, {cb}. In these six arrangements the order of the elements matters. Two self-evident principles are at work in the enumeration of sets:

The addition rule: if one event occurs in n ways and another event in m ways then there are n + m ways in which *either* of these two events can occur. For example, if the events consist of selecting from two sets of elements {a, b, c, d} and {A, B} there are 4 + 2 = 6 ways of selecting one element from either of these sets.

E. Beltrami, *Digressions in Elementary Probability*, Quantitative Methods in the Humanities and Social Sciences, https://doi.org/10.1007/978-3-031-70051-4

The product rule: if one event can occur in n ways and a second event in m ways, there are n×m ways *both* of these two events can occur. Using the same two sets {a, b, c, d} and {A, B} considered above, there are 4 × 2 = 8 ways of selecting two elements, one from the first set and one from the second. Certain identities hold:

$$P(n, k) = P(k, k)C(n, k) \tag{1}$$

This is because an ordered arrangement of k objects out of n can be done by first selecting k objects and then arranging these k objects in order. The product rule was used here. Next,

$$C(n, k) = C(n - 1, k - 1) + C(n - 1, k) \tag{2}$$

This is seen by first marking one of the distinct objects. The number of ways of selecting k objects so that the marked one is always included is C(n − 1, k − 1), and the number of ways of selecting k objects that do not include the marked object is C (n − 1, k). Here we used the sum rule.

To obtain an expression for P(n, k) note that there are n ways to fill the first position of a distinct ordering of k objects, n − 1 ways to fill the second position (since one has already been chosen), and n − k + 1 ways to fill the last position. Therefore, according to the product rule,

$$P(n, k) = n(n - 1) \ldots (n - 1 + k) = n!/(n - k)! = n(n - 1) \ldots (n–k + 1) \tag{3}$$

It is now easy to get an expression for C(n, k). From identity (1) we see that

$$C(n, k) = P(n, k)/P(k, k) = n!/(n - k)!k! \tag{4}$$

It follows immediately from (4) that C(n, k) = C(n, n − k). Note that C(n, n) = 1 and C(n, 1) = n. Also, C(n, 0) = C(n, n − 0) = 1. A central result is the *Binomial Theorem,* namely:

$$(a + b)^n = \sum_{k=0}^{n} C(n, k)a^k b^{n-k}$$

Proof We use induction. This identity is evidently true for n = 1. Assume it to be true for n − 1. Now $(a + b)^n = a(a + b)^{n-1} + b(a + b)^{n-1}$ where, by assumption, $(a + b)^{n-1} = \sum C(n - 1, k)a^k b^{n-1-k}$, summed from 0 to n − 1. Using the identity C(n, k) = C(n−1, k−1) + C(n−1, k) and, as a special case of this same identity, that C(n−1, n) = C(n, n) − C(n−1, n−1) = 0, the induction goes through and the theorem is established.

Probability Spaces

A set S is called the *sample space* if it consists of a *finite or countably infinite* number of possible outcomes of some hypothetical experiment, probe, or trial. Subsets of S are called *events.*

Every element of S, namely one of the possible outcomes of the experiment or trial, is assigned a probability. In the present context, this is taken to mean some degree of belief in the likelihood that a particular outcome actually takes place. The assumption is that each outcome is due to chance and is inherently uncertain. The sum of the probabilities over each of the possible outcomes, namely the entire sample space, is taken to be unity. That is, P(S) = 1, where P denotes "probability of". In general, for any event A, $0 \leq P(A) \leq 1$

A typical and quite simple illustration is that of tossing two coins in succession. The sample space consists of four possible outcomes, namely HH, HT, TH, TT where H denotes "heads" and T "tails". If we believe that the coin toss is unbiased and free of any contrivance that would cause H to be more or less likely than T, then it is reasonable to assign the equal probabilities of 1/4 to each of these mutually exclusive outcomes. The sample space is the union of the four possibilities and its measure is unity, as required.

Since an event A is a subset of the probable outcomes, the probability of A is the sum of the probabilities of the elements contained in A. In the previous example an event might be that at least one head occurs in two tosses and this consists of three possibilities out of four total: HH, HT, TH. The probability of this event is accordingly 3/4.

Given event A let ~A denote the *complement* of A in S so that P(~A) = 1 − P(A). Also, if A and B are two events we say they are *disjoint* if $P(A \cup B) = P(A) + P(B)$ where $A \cup B$ denotes the union of A and B.

Given a sample space S and an event $B \subset S$ that is known to have occurred, all future events must now be computed relative to B. That is, B is now the new sample space and the only part of an event A that is relevant is $A \cap B$ where $\cap$ denotes the intersection of A and B. Defining the *conditional probability of A given B,* written as P(A|B), as a quantity proportional to $P(A \cap B)$, it is easy to see that the constant of proportionality must be 1/P(B) in order that P(B|B) equal one. Thus,

$$P(A|B) = P(A \cap B)/P(B) \tag{5a}$$

or, equivalently, since it is also true that $P(B|A) = P(A \cap B)/P(A)$,

$$P(A|B) = P(B|A)P(A)/P(B) \tag{5b}$$

and this is known as *Baye's Theorem.*

In the preceding example, suppose we are told that in two tosses at least one of them was a head. This narrows the sample space to the event B consisting of HH, HT, TH since TT is no longer a possibility. Let A be the event that the other toss also was a head, namely that HH took place. The conditional probability of A given B is now 1/3. Conditioning is a tricky notion that easily leads to unanticipated results.

Sometimes it is convenient to write the denominator in (5a and 5b) in a different way. To do this note that the sample space S can be written as a union $S = A \cup \sim A$ and so

$$B = B \cap S = B \cap (A \cup \sim A) = (B \cap A) \cup (B \cap \sim A).$$

This last identity is easily established by showing that each side is included in the other. Since $A \cap B$ and $A \cap \sim B$ are disjoint events, it now follows that $P(B) = P(A \cap B) + P(A \cap \sim B)$ and, using (5a and 5b), we obtain

$$P(B) = P(B|A)P(A) + P(B|\sim A)P(\sim A) \tag{6}$$

which is sometimes referred too as the *Law of Total Probability*. We will come across this relation again later in the tutorial using a different argument.

Event A is said to be *independent* of event B if the conditional probability of A, given B, is the probability of A itself, which paraphrases the idea that the knowledge of B having occurred has no influence on the likelihood of A taking place. From (5a and 5b) we see that A is independent of B if and only if

$$P(A \cap B) = P(A)P(B) \tag{7}$$

Random Variables

A function X defined on a sample space S assigns a non−negative real number to each outcome in S and is called a *random variable* since its values are determined by the probabilistic outcomes of the elements x of S. In the simplest case suppose that S has just two elements labelled as "success" and "failure" and that $X = 1$ when success occurs and $X = 0$ when failure happens. For another example let S be the space of four outcomes of two tosses of a fair coin that was considered earlier: $S = \{HH, HT, TH, TT\}$ and define X to be the number of heads in the two tosses. Then X takes on the values 0, 1, 2.

X can take on a finite or countably infinite number of values and we define the *probability mass function* of X by $p(a) = P(X = a)$ where a is any of the values that X assumes on S. That is, $X(x) = a$. In the first example of the previous paragraph, $p(1) = P(X = 1) = p$ and $p(0) = P(X = 0) = q = 1 - p$ where $0 < p < 1$ (p is the probability of "success"). In this case X is known as a *Bernoulli random variable*. In the second example, $p(0) = P(X = 0) = 1/4$ since the probability of no heads occurs in only one case, namely TT. Then $p(1) = P(X = 1) = 1/2$ since the probability one head in two tosses happens in two ways: HT and TH. Finally $p(2) = P(X = 2)$ is again 1/4 since this event corresponds to HH.

An important random variable in applications, designated by S_n, is defined on the space S consisting of the outcomes of n *independent* trials each of which is either a success with probability p or a failure with probability q = 1 − p. The S_n is the number of successes in the n trials which can also be thought of as a sum of n Bernoulli random variables, one for each trials. Evidently S_n, known as a *Binomial random variable*, can take on the values k = 0, 1, 2, . . ., n and its probability mass function is $p(k) = P(S_n = k) = C(n, j)p^k q^{n-k}$. The reason for this expression is that the number of ways k success can be arranged among n trials is C(n, k) and, because of independence, each of these events has a probability of k successes and n = k failures given by $p^k q^{n-k}$.

Expected Values

A random variable X has an expected or average value E(X) that is a weighted average of the possible values it can take on, weighted by the probability of it assuming any particular value. That is, $E(X) = \sum x\, p(x)$ summed over all the values x that X assumes. Thus, for example, for a Bernoulli variable E(X) = 1 times p(1) + 0 times p(0) = p whereas for the two toss of a coin example

$$E(X) = 0 \text{ times } p(TT) + 1 \text{ times } p(HT \cup TH) + 2 \text{ times } p(HH) = 1/2 + 1/2 = 1..$$

To make further progress I now consider two random variables X and Y defined on S and define the *joint probability mass function* by p(x, y) = P(X = x, Y = y) where x and y are values separately taken on by X and Y. The *marginal mass function* of X is obtained from p(x, y) as $p_X(x) = \sum p(x, y)$ summed over all possible y values. A similar definition gives rise to the marginal mass function of Y.

I can now show that E(X + Y) = E(X) + E(Y). This follows since E(X) + E (Y) $= \sum xp(x) + \sum yp(y) = \sum x \sum p(x, y) + \sum y \sum p(x, y)$ in which the first double sum is over x first and y second with just the opposite for the second double sum. Adding these sums gives $\sum \sum (x + y)p(x, y) = E(X + Y)$. A simple inductive argument now establishes that if $\sum X_k$ is a sum of random variables on S then $E(\sum X_k) = \sum E(X_k)$. One can then show that $E(S_n) = np$ since S_n is a sum of n Bernoulli random variables.

Let g(X) be a real valued function defined on the range of values of some random variable X: since X maps the sample space S into a set S' by a =X(x) then g(X) is a random variable defined by g(a) for all values a in S'. It can be shown (Ross [113]) that $E(g(X)) = \sum g(x)p(x)$ where p(x) is the mass function of X. Thus, for example, E(cX) = cE(X) for any constant c and $E(X^2) = \sum x^2 p(x)$.

The *variance* of X is defined to be the expected value of the squared difference between X and its expected value E(X). It is a measure of the discrepancy between a random variable and its average, namely the spread of values of X about its average:

$$\begin{aligned} \mathrm{Var}(X) &= E(X - E(X))^2 \\ &= \sum (x - E(X))^2 p(x) \text{summed over all values of X.} \end{aligned} \quad (8)$$

Since $(x - E(X))^2 = x^2 - 2xE(X) + E(X)^2$ and, noting the fact that E(X) is a constant and $E(xE(X)) = E(X)^2$, once finds that

$$\mathrm{Var}(X) = E(X^2) - E(X)^2 \quad (9)$$

An example of (9) would be the variance of a Bernoulli random variable in which $E(X) = p$ and $E(X^2)$ is also p since the square of one is still one. But $E(x)^2 = p^2$ and so $\mathrm{Var}(X) = p - p^2 = pq$. The variance of S_n will be established later in this tutorial. Note that $\mathrm{Var}(cX) = c^2\mathrm{Var}(X)$ for any constant c.

To illustrate some of the contents of the previous sections of this tutorial we will introduce a seemingly paradoxical calculation that illustrates the sometimes confounding nature of conditional probability. Other examples appear throughout this book and they add a certain piquancy to the study of elementary probability:

Paradox of the Second Ace

A bridge hand is a random selection of 13 cards from a deck of 52 cards and so a total of C(52, 13) different hands can be dealt to four players. There are four aces in the deck. Suppose a player says "I have an ace:" Call this event B and let event A be "there are at least two aces in the hand". We want to find P(A). To begin, there are C (48, 13) hands with no ace (the four aces were removed from the deck) and so the number of hands with at least one ace is C(52, 13) – C(48, 13) which constitutes event B.

What about having at least two aces? We get this by subtracting out those hands with exactly one ace. To do this first remove all aces from the deck and deal a hand of 12 cards (having no aces). This can be done in C(48, 12) ways. Next add back in a single ace to each hand. There are four ways of doing this, one for each ace, and so the event $A \cap B$ (two or more aces) occurs in C(52, 13) – C(48, 13) – 4 *times* C(48, 12) ways. By dividing the number of ways each event occurs by the total number of possible hands, namely C(52, 13), one finds that P(A) equals.

$$1 - 4 \text{ times } (C(48, 12)/[C(52, 13) - C(48, 13)]) < 1/2.$$

Now suppose the player instead says " I have an ace of hearts". Remove this particular ace from the deck to deal a hand with 12 cards. The number of hands without this ace is now C(51, 12) and, as before, if we now add the ace of hearts back into a hand (only one way to do this) it follows that C(51, 12) is the total number of hands containing at least the ace of hearts. Call this event B. To find the number of hands with the ace of hearts and at least one other ace, event $A \cap B$, subtract all hands with no ace. There are C(48, 12) of these but now, instead of re-inserting an ace back into a hand in one of four different ways as before we are restricted to restoring only the ace of hearts (there is one way of doing this) because our sample space has shrunken to exclude any hand that doesn't include an ace of hearts. This gives C(51, 12) – C(48, 12) hands containing at least one ace besides the ace of hearts. Note that event A has a slightly different meaning here than previously.

Given that we know that someone declared an ace of hearts it follow that there are only C(51, 12) possible hands to consider as a sample space and therefore P(A) equals.

$$1 = C(48, 12)/C(51, 12) > 1/2$$

So this last probability is *not equal* to the one computed previously! We have a surprising and seemingly absurd result that by announcing the suit of the ace we hold one gets a different probability for two or more aces then when someone simply announces that they hold one of four possible aces.

To make this conclusion a bit more intuitive consider a special case of a deck of just 4 cards, an ace of hearts and one of spades labelled A1 and A2 respectively and two other cards, say deuces, designated as D1 and D2. A hand consists of just two cards and so there are C(4, 2) = 6 possible hands altogether. It is straightforward that 5 of the hands contain at; east one ace since only the (D1, D2) hand is excluded. But only one of the five hands has two aces and so the probability of two aces is 1/5. Now suppose that we know that a hand contains an ace of hearts. There are C(2,1) = three such possibilities: (A1, A2), (A1, D1), (A1, D2). It follows that the probability of a hand with two aces is now 1/3. Once again the conditional information has altered the likelihoods.

An excellent account of very similar problems involving tricky interpretations of conditional probabilities can be found in a paper by Bar-Hillel and Falk [73].

Conditional Expectation

Let X, Y be discrete jointly distributed random variables with p(x,y) = P(X = x, Y = y) having marginal probabilities $p_X(x) = p(X = x) = \sum p(x, y)$ summed over the discrete set of y values.

One defines the *independence of X and Y* over S by $p(x, y) = p_X(x)p_Y(y)$ in keeping with the independence of two events in S as described earlier. It follows that when X and Y are independent that $E(XY) = \sum \sum xyp(x, y) = \sum \sum xp_X(x)$-$p_Y(y) = \sum xp_X(x) \sum yp_Y(y) = E(X)E(Y)$ where the outer sum is over x values and the inner sum over y values.

The conditional probability of X given Y = y is defined by $p_{X\|Y}(x\|y) = P(X = x\|Y = y) = p(x, y)/p_Y(y)$ when the denominator is non-zero. Then the *conditional expectation of X given Y = y* is defined by $(X \| Y = y) = \sum x\, p_{X\|Y}(x \| y)$. It is useful to establish the following theorem

$$E(X) = E\left(E\left(X \| Y\right)\right)$$

To prove this we need to establish that

$$E(X) = \sum E\left(X \| Y = y\right) p(Y = y) \quad \text{summed over y values} \tag{10}$$

The right side of (10) can be expressed as $\sum\sum x\, P(X = x \| Y = y)P(Y = y)$ where the first sum is now over x values and the other sum over y as before. But this equals

$$\sum\sum x[p(X = x, Y = y)/p(Y = y)]\, p(Y = y) = \sum\sum x\, p(X = x, Y = y)$$

and, in turn, equals $\sum x \sum p(X = x, Y = y) = \sum x\, p(X = x)$ summed over x and this is E(X).

The theorem is now used to compute probabilities. Let A be an event with probability P(A) and suppose that X is a random variable defined by X = 1 if A occurs and 0 otherwise. It is straightforward that E(X) = P(A) and that $E(X \| Y = y) = P(A \| Y = y)$ for any random variable Y. Because of relation (10) in the above theorem, we get

$$P(A) = E\left[\left(P(A) \| Y\right)\right] = \sum P\left(A \| Y = y\right) P(Y = y) \tag{11}$$

where the sum is over y values, a result used in the fourth section of the book in which we discuss a subtly biased estimation.

If Y takes on the values $y_1, y_2, \ldots, y_n$ and if B_k are defined as disjoint events $B_k = (Y = y_k)$ relation (11) becomes

$$P(A) = \sum P\left(A \| B_k\right) P(B_k) \tag{12}$$

summed from 1 to n. Of course (12) can also be obtained by noting that $A = \cup (A \cap B_k)$ so that $P(A) = \sum P(A \cap B_k) = \sum P(A \| B_k)P(B_k)$. This is, once again, the Law of Total Probability derived earlier in a different manner.

Covariance

An extension of the idea of variance appears in a few places in the book especially in the discussion of polls when evaluating two candidates. The Covariance of random variables X and Y is defined by $Cov(X, Y) = E[(X - E(X))(Y - E(Y))]$ and expanding what appears within brackets this becomes E(XY) − E(X)E(Y). When X and Y are independent then evidently Cov(X,Y) = 0.

It is immediate that Cov(X, Y) = Cov(Y, X) and Cov(X, X) = Var(X). It is also not difficult to see that if one has a sum of random variables $\sum X_j$ and $\sum Y_k$, summed up to n in one case and up to m in the other, then $\text{Cov}(\sum X_j, \sum Y_k) = \sum\sum \text{Cov}(X_j, Y_k)$. It now follows that $\text{Var}(\sum X_j) = \text{Cov}(\sum X_j, \sum X_k) = \sum\sum \text{Cov}(X_j, X_k) = \sum \text{Var}(X_j) + \sum\sum \text{Cov}(X_j, X_k)$ in which the double sum is over all j and k for which $j \neq k$. Note, incidentally, that this double sum is also equal to twice the double sum in which $j < k$, an evident symmetry.

If $X_1, \ldots, X_n$ are independent variables then $\text{Cov}(X_j, X_k) = 0$ for all $j \neq k$ and in this case one sees that $\text{Var}(\sum X_j) = \sum \text{Var}(X_j)$. An example of this is when the X_j are independent Bernoulli random variables for which one now obtains that $\text{Var}(S_n) = \text{Var}(\sum X_j) = npq$.

The Law of Large Numbers

A core result, a centerpiece of probability theory, is the Law of Large Numbers which appears both implicitly and explicitly in many sections of the book. A modest version of this theorem, the weak law, is fairly easy to prove but more apt for our needs is the so-called "strong law" which requires a detailed argument that is somewhat beyond this book since our focus is on *elementary ideas in probability*. But what it states and, more especially, what it implies, is important. I'll state it in terms of Bernoulli trials, ignoring a technical caveat that is not important for our purposes.

Suppose an event A has a probability p = P(A) of occurring and let X_k be a Bernoulli random variable to equal 1 when A occurs indicating 'success' and is 0 otherwise. Then the binomial variable $S_n = \sum X_k$ counts the number of successes in n trials or n repetitions of an experiment or observation and S_n/n, the *sample mean*, is the frequency or mean number of successes in these trials. What the Law of Large Numbers effectively tells us is that as n increases the sample means are certain to approach p. Even if p is initially unknown one can obtain better and better approximations to it simply by increasing the sample size n. In applications it is often the case that a modest value of n already give a good empirical estimate of P(A) as the number m of successes divided by the number n of trials: p = m/n.

This theorem was first established by Jacob Bernoulli at the beginning of the eighteenth century and he illustrated it by an example of an urn containing 300 white balls and 200 black balls so that the probability p of drawing a white ball ("success") at random in a single selection is 3/5. Begin by assuming that one does not know the distribution of white and black balls. Bernoulli then samples the urn repeatedly with replacement and obtains a frequency of white balls roughly equal to 3/5.

However, by the middle of the eighteenth century a different slant on Bernoulli's frequentist approach was posthumously proposed by Thomas Bayes that assigned an a-priori degree of belief on the likelihood of a white ball or, more generally, on an event A. As evidence becomes available that is compatible with this belief one can update the subjective a-priori estimate using what we earlier called Baye's Theorem. This is an "inverse probability" because it infers backward from evidence to cause, from observation to hypothesis, from an objective viewpoint into one that is now subjective. Bernoulli computes the probabilities of frequencies given a chance event while Bayes computes the probability of the chance event given the frequencies. If E is evidence and C the chanced event then Bernoulli obtains the conditional probability P(E||C) in contrast to P(C||E).

Here is another example of how conditional probabilities are confounding. along with the others we've encountered in a number of vignettes. Amanda Knox, an exchange student from Seattle, was accused in 2007 of murdering her roommate in Perugia, Italy. A murder conviction in 2009 was invalidated in 2011. One piece of evidence was a knife collected from the victim's apartment containing a tiny trace of DNA from the victim, but it was inconclusive. In 2013 the trial was reopened but the judge refused to have another test done on the original blood sample since the first one was questionable. As reported in a NY Times article [68] that was a mistake because if the second test gave the same result this new data would increase the probability of guilt. *That is what Baye's rule tells us*. It's like tossing a coin ten times and find it comes up heads nine or ten times. You suspect it could be biased. If you now repeat for another ten tosses and you still get nine or ten heads this adds to your conviction the coin is indeed biased. As the author's of the article state "The judges rejection of a retest based on the notion that a confirming retest would tell us nothing about the likelihood that the DNA was a match was a serious error."

Another illustration concerns ESP experiments that support the idea that ESP is a real effect. But, as reported in another NY Times article by Carey [67] "Claims that defy almost every law of science are by definition extraordinary and thus require extraordinary evidence." In this case the prior odds in favor of ESP are small and there is reason to be skeptical of the evidence.

There are several places in which I use a *different measure of uncertainty known as odds* which is the number of successes divided by the number n − m of failures: odds = m/(n − m) in contrast the empirical notion of probability given, as we saw, by the number of successes divided by the number of trials, m/n. Evidently one can go from one measure to the other. For example odds of 3: 2, usually stated as odds of 3 to 2 ("three favorable outcomes against two unfavorable"), corresponds to a probability of 3/5 since n = 2 +3 or, for another example, odds of 2 to 1 is the same as a probability of 2/3.

In this book Baye's Theorem, as given by equation 5b generally rewritten in odds form in part because the probability of event B conveniently drops out when we form the *likelihood function* P(B||A)/P(B||~A) where ~A means "not A", namely the complement of A in S.

There is one random variable that also appears in the book, namely the *normal random variable* which varies continuously rather than discretely. There is an extensive theoretical background regarding the normal variable leading to another fundamental theorem of probability known as the *Central Limit Theorem* which is vastly outside the scope of this tutorial. Fortunately we require a very limited special case for (almost all) the applications that are presented in the various vignettes and this can be succinctly summarized by stating that although we know that S_n/n approaches p as n gets larger it is possible to give a more precise estimate:

Let σ, known to statisticians as a *standard deviation*, denote the square root of variance. Then, in particular, the standard deviation of S_n is the square root of npq and so the standard deviation σ of the sample mean S_n/n becomes $\sqrt{(pq/n)}$. The Central Limit Theorem then tells us that as n gets larger one can compute, for any constant a, an increasingly better estimate of the probability that $-a\sigma < S_n/n - p < a\sigma$. In the special case in which $a = 2$ (more accurately $a = 1.96$) and $p = q = 1/2$ then, since σ is now (close to) $1/\sqrt{n}$, we find that is close to .95 when n is large enough. Again, in practice, n doesn't have to be that large for this estimate to hold. In case $p \neq 1/2$ one needs to have available an estimate for σ and in this situation what one gets is that $P(-2\sigma < S_n/n - p < 2\sigma)$ is approximately. 95.

Since the normal random variable also appears in the seventh vignette I'll review here the salient facts without proof since they require a modest amount of elementary calculus and this can be found in Ross [13].

A random variable X is continuous if there is a continuous function f(x), the *density function of X*, for which P(X takes on values in a subset B of the real numbers) equals the integral of f(x) over the set B. This corresponds, for a discrete random variable, to a sum of values of a probability mass function. In particular X is a *normal random variable* with parameters θ and σ^2 if its density function is given by

$$f(x) = 1/\sigma\sqrt{2\pi} \text{ times } \exp\left(-(x-\theta)^2/2\sigma^2\right) \tag{13}$$

One then shows that the integral $\int f(x)dx$ over all real numbers equals one, as it should to be a proper probability measure, and also that $E(X) = \theta$ and $Var(X) = \sigma^2$. The normal density (13) is the familiar bell-shaped curve centered at its maximum value where $x = \theta$ and having a spread about θ measured by σ^2. If one defines a random variable

Z as $(X - \theta)/\sigma$ then it turns out that Z has a normal density and it is readily verified that in this case Z has mean zero and variance one so that $P(-a < Z < a)$ is the integral of $1/\sqrt{2\pi} \exp(-x^2)$ between $-a$ and $+a$. When $a = 1.96$ this probability is .95. This applies, in particular, when $Z = (S_n/n - p)/\sigma$ and this is the source for the 95% confidence interval that I introduce within the book and in the previous paragraph.

Another approximation to S_n is developed in the second vignette in terms of a *Poisson random variable* X which takes on the values $k = 0, 1, 2, \ldots$. Its mass function is $p(k, \lambda) = \lambda^k \exp(-\lambda)/k!$ in which λ is the expected value $E(X) = \sum kp(k, \lambda)$ summed from zero to infinity. To see this note that $\sum kp(k, \lambda) = \lambda \exp(-\lambda) \sum \lambda^{k-1}/(k-1)!$ now summed from 1 to infinity. Letting $m = k - 1$ the sum becomes $\sum \lambda^m/m!$ with m ranging from 0 to infinity. You may recall from elementary calculus the last sum is the series expansion for $\exp(\lambda)$. The two exponentials then cancel verifying that $E(X) = \lambda$ as stated.

A quite similar computation, using $\text{Var}(X) = E(X^2) - E(X)^2$, shows that Var (X) is also λ.

A fairly straightforward proof (see Ross [13]) establishes that $P(S_n = k)$ is closely approximated by $p(k, \lambda)$ when n is large enough (again, it is often true that n need not be all that large) assuming p is small enough so that np is moderate in value.

Appendix: Matlab Codes

The following m.files were written by the author. No doubt there are more succinct and elegant ways of coding them, but they do work.

```
3/30/24 1:22 PM  /Users/edbeltrami/Document.../berresford.m   1 of 1

% Berresford

n=input('number of coin tosses .   ');
k=input('minimum size of a run .   ');
p=input('probability of a head .    ');
b=zeros(1,n);
r=p^k;
for j=1:(k-1)
    b(j)=0;
    end

b(k)=r;
b(k+1)=r*(2-p);
for s=(k+2):n
    b(s)=b(s-1)+((1-b(s-k-1))*r*(1-p));

  end
    b(n)

%fprintf('prob of runs of heads of size no less than k   :%g.\n',c(n))
```


E. Beltrami, *Digressions in Elementary Probability*, Quantitative Methods in the Humanities and Social Sciences, https://doi.org/10.1007/978-3-031-70051-4

Computes the probability of a success run of length at least k in n Bernoulli trials

5/7/22 2:08 PM /Users/edbeltrami/Documen.../millersingle.m 1 of 1

```
% TO COMPUTE THE PROBABILITY THAT A HEAD FLIP FOLLOWS A HEAD IN A SEDQUENCE
% OF COIN TOSSES USING MS'S PAPER

n=input('number of repetitions .     ');
m=input('length of binary string .     ');
p=input('probability of a head .   ');

u=[];,hh=[];,L=[];h=[];

% COMPUTE A RUN OF COIN TOSSES WITH PROBABILITY P OF A HEAD

for j=1:n
    for k=1:m
        if rand<p
            s=1;
        else s=0;end;
        u=[u,s];
    end
        for j=1:(m-1)
            if u(j)==1
                h=[h,1];
            else h=[h,0]; end
        end
        for i=1:(m-1)
            if sum(u(i:i+1))==2
                hh=[hh,1];
            else hh=[hh,0];end
        end
            if nnz(h)>0

            L=[L,(sum(hh)/sum(h))];
            end
            u=[];h=[];hh=[];
end
            mean(L)
```

Computes the probability that a head flip follows a head in a sequence of coin tosses

10/21/23 2:53 PM /Users/edbeltrami/Docume.../negbinomial.m 1 of 1

```
%negbinomial. Lep p be prob of success in a sequence of Bernoulli
%trials. Continue until there are k successes. If X is the number of trials
%before k successes, with X=k, k+1, ... then the prob(X = r) is given by
%the negative binomial
%p=input('probability of success      ');
n=input('maximum number of trials      ');
k=input('required number of successes      ');
q=1-p;m=x-k;
u=[];
for x=k:n
    y=nchoosek(x-1,k-1)/2^x;
   u=[u;y];
   z=k:n;
end
    s=sum(u)
    %s is the total prob of seeing k successes in n trfials
    bar(z,u)
    xlabel('number of trials for k successes')
    ylabel('prob of the number of trials before k successes')
```

Computes the negative binomial density

```
C:\MATLAB6p1\work\Montmort.m                                          Page 1
January 29, 2010                                                  4:12:55 PM

% montmort.m
% code to compute the probability of matching k objects out of n
% for k = 0, 1, ..., n and also the average number of matched objects.
%
n=input('total number of items    ');
k=input('number of items to match    ');
gg=[];ss=[];tt=[];
for s=0:n
    for j=0:n-s
    u=((-1)^j)/gamma(j+1);
    ss=[ss,u];
w=sum(ss)/gamma(s+1);
end
ss=[];
tt=[tt,w];
gg=[gg,s];
end
fprintf('the average number of matched objects:%g.\n',sum(gg.*tt))
%
k
t=k+1;
%
fprintf('probability that k objects match out of n:%g.\n',tt(t))
```

Computes the probability of exact k matches and the average for n items.

```
C:\MATLAB6p1\work\cointoss.m                                          Page 1
January 29, 2010                                                  2:57:54 PM

% cointoss.m
% code to simulate fluctuations in accumulated gains in a coin-tossing
% experiment It computes the average gain, in contrast to accumulated gain.
% It also outputs a computer-generated graphic
% illustrating the fluctuations of wins and losses in tosses of a fair
% coin. The number of trials are in multiples of 10.
%
% by calling cointoss in the MATLAB workspace results in a prompt for the
% number of trials one wishes to simulate.
%
ss=[];feller=[];
n=input('number of trials is...');
for i=1:n
if rand<.5
s=1;
else s=-1;
end
ss=[ss,s];
gain=sum(ss);
feller=[feller,gain];
end
z=feller;
plot(feller(1:10:n))
xlabel('number of coin tosses in multiples of ten')
ylabel('fluctuation of wins and losses')
% we now count how many times gain is positive in n trials:
for k=1:n
if feller(k)<0
feller(k)=0;else feller(k);end
end
fprintf('the number of times one player is in the lead: %g.\n',nnz(feller))
fprintf('the average gain of a player: %g.\n',mean(ss))
```

Computes and displays the fluctuations in a game of heads and tails.

```
C:\MATLAB6p1\work\parrondo.m                                    Page 1
July 29, 2008                                               2:24:42 PM

% Parrondo's Game
d=input('initial capital    ');
win=[];
for j=1:2000
    for i=1:100
        if rand<.5
            if rand<.495
            d=d+1;
        else
            d=d-1;
        end
    else
        if rem(d,3)==0
            if rand <.095
                d=d+1;
            else
                d=d-1;
            end
        else
            if rand<.745
                d=d+1;
            else
                d=d-1;
            end
        end
    end
end
win=[win;d];
end
plot(win)
title('parrondos game')
```

Simulates Parrondo's game and displays the accumulated gains.

```
3/30/24 11:39 AM   /Users/edbeltrami/Documents/.../shannon.m    1 of 1

% Shannon.m
% Shannon's Demon for accumulated earningsusing rebalanced ninvestments
% (solid mline) versus conventional investment (dashed line)
h1=input('cash .    ');
h2=input('stocks .       ');
s1=input('cash .      ');
s2=input('stocks .       ');
n=input('number of trades .     ');
H=[];S=[];
for k=1:n

    if rand<.5
        h2=h2/2;
        s2=s2/2;c=(s1+s2)/2;
        s1=c;s2=c;
    else h2=2*h2;
        s2=2*s2;c=(s1+s2)/2;
        s1=c;s2=c;
    end
    H=[H,h2];S=[S,s2];
end
    semilogy(H(1:n))
    hold on
   semilogy(S(1:n))

    xlabel('nimber of trades')
    ylabel('accumulated earnings')
    title('demon wealth')
```

Computes Shannon's Demon, an investment strategy, and displays the accumulated gains.

5/1/24 3:09 PM /Users/edbeltrami/Documents/MA.../poisson.m 1 of 1

```
% poisson
% generates cumulative poisson arrivals at mean rate lambda.
% and also the exponentially distributed inter-arrival times
n=input('number of arrivals    ');
lambda=input('value of lambda    ');
p=[];z=[];q=[];
for i=1:n
    v=rand;
    u=-log(v)/lambda;
    q=[q,u];
    p=sum([p,u]);
    z=[z,p];
end
'The exponential inter-arrival times:'
q
'The cumulative sum of Poisson arrival times:'
z
```

Computes exponential inter-arrival times from a Poisson distribution and their cumulative sum.

References

1. Bar-Hillel, Maya, and Willem Wagenaar, The Perception of Randomness, Adv. Applied Math. 12, 428–454, 1991
2. Falk, Ruma, and Clifford Konold, Making Sense of Randomness, Psychological Review 104, 301–318, 1997.
3. Beltrami, Edward, What is Random?, 2nd Edition, Springer, 2020
4. Feller, William, Probability Theory, Volume 1, 2nd Edition, Wiley, 1957
5. Miller, Joshua and Adam Sanjurjo, Surprised by the Gambler's and Hot Hands Fallacies? A Truth in the Law of Small Numbers, Econometrica 86, 2019–2047, 2018
6. Miller, Joshua, and Adam Sanjurjo, Bridge from Monty Hall to the Hot Hand: Restricted Choice, Selection Bias, and Empirical Practice, J. Econ. Perspect 33, 144–162, 2019
7. Tversky, Amos, and Thomas Gilovich, The Cold Facts about the "Hot Hand" in Basketball, Chance 2, 16–21, 1989.
8. Gilovich, Thomas, Vallone, Robert and Amos Tversky, The Hot Hand in Basketball: On the Misconception of Random Sequences, Cognitive Psychology 17, 295–314, 1985.
9. Bar-Hillel, Maya, and Yosef Rinott, Comments on a Hot Hand Paper by Miller and Sanjurjo, SSSN Electronic Journal, Jan. 2015
10. Gusfield, Dan, Hot Hands, Streaks and Coin-Flips: Numerical Nonsense in The New York Times, arXiv: 1512–08773, Dec. 29, 2015
11. Bloom, David, Singles in a Sequence of Coin Tosses, Coll. Math. J. 29, 120–127, 1998
12. Berresford, Geoffrey, Runs in Coin Tossing: Randomness Revealed, Coll. Math. J. 33, 391–394, 2002
13. Ross, Sheldon, A First Course in Probability Theory, 8th Edition, Prentice Hall, 2010
14. Beltrami, Edward, and Jay Mendelsohn, More Thoughts Regarding Di Maggio's 1941 Streak, Baseball Res. J., 31–34, 2010
15. Miller, Joshua, and Adam Sanjurjo, Momentum Isn't Magic, Vindicating the Hot Hand with the Mathematics of Streaks, Sci. Am 2018 (reprinted from The Conversation 2017)
16. Wardrop, Robert, Simpson's Paradox and the Hot Hand in Basketball, Am. Statistician 49, 24–28, 1995.
17. Reynolds, Gretchen, Are Hot Hands in Sports the Real Thing?, NY Times, May 8, 2013.
18. Gawande, Atul, *The Cancer Cluster Myth*, The New Yorker, February 8, 1999
19. Rosenhouse, Jason, The Monty Hall Problem, Oxford, 2009
20. Dorsey-Palmateer, Reid, and Gary Smith, Bowlers' hot hands, American Statistician 58, 38–45, 2004.
21. Wainer, Hubert, The Most Dangerous Equation, American Scientist 95, 249–256, 2007

E. Beltrami, *Digressions in Elementary Probability*, Quantitative Methods in the Humanities and Social Sciences, https://doi.org/10.1007/978-3-031-70051-4

22. Beltrami, Edward, Mathematical Models for the Social and Biological Sciences, Academic Press, 2003
23. Hubert, Lawrence, and Hubert Wainer, A Statistical Guide for the Ethically Perplexed, CRC Press, 2013
24. Luria, Salvador, and Max Delbruck, Mutation of Bacteria from Virus Sensitivity to Virus Resistance, Genetics 28, 491–511, 1943
25. Use of Colonoscopy to Screen Asymptomatic Adults, New England Journal of Medicine 343, 2000, 162–208.
26. Grady, Denise, Broader, and More Expensive, Test Needed to Spot Colon Cancers, Studies Say, NY Times, Sept. 20,, 2000.
27. Tu, Yu-Kang, and Mark Gilthorpe, The Most Dangerous Hospital or the Most Dangerous Equation?, BMC Health Services Research, 1–5, 2007
28. Gigerenzer, Gerd, Calculated Risks, Simon & Schuster, 2002
29. Franklin, Charles, The Margin of Error for Differences in Polls, Marquette Law School publication, 2007
30. Abramson, Morton, and Jurgen Moser, More Birthday Surprises, Am. Math Monthly 77,1970, 856–858
31. Cohen, Jack, and Ian Stewart, That's Amazing, Isn't It?, New Scientist, January 17, 1998
32. Falk, Ruma, Judgement of Coincidences, Mine Versus Yours, American Journal of Psychology 102, 1989, 477–493
33. Gelman Andrew, and Joshua Miller, Laplaces's theories of cognitive illusions, heuristics, and biases, online paper, December 1, 2018.
34. Stigler, Stephen, Regression Towards the Mean, Historically Considered, Statistical Methods in Medical Research 6, 1997, 103–114
35. Ellenberg, Jordan, How Not to be Wrong, Penguin, 2014
36. Kahneman, Daniel, Thinking, Fast and Slow, Farrar, Straus and Giroux, 2011.
37. Kahn, David, The Code Breakers, Macmillan, 1967
38. Simpson, Edward, Bayes at Bletchley Park, Significance, 2010, 76–80
39. Zabell, Sandy, Commentary on Alan Turing: the Application of Probability to Cryptography, Cryptologia 36, 191–214
40. Cavett, Dick, Seriously, What Are The Odds?, NY Times, May 8, 2009.
41. Tierney, John, Behind Monty Hall's Doors: Puzzle, Debate, Answer?, NY Times, July 21, 1991.
42. Poundstone, William, Fortune's Formula, Hill and Wang, 2005.
43. Kolesar, Peter, and Edward Blum, Square-Root Laws for Fire Engine Response Distances, Management Sciences 19, 1973, 1368–1378.
44. Efron, Bradley, and Carl Morris, Stein's Paradox in Statistics, Scientific American, May, 1977.
45. Birnbaum, Phillip, Can Estimates of Wheat Production Improve our Evaluation of Baseball Talent?, Sabermetric Research, Oct 31, 2010, 1–10
46. Stein's Paradox, Wikipedia
47. Kolata, Gina, Cholesterol-Slashing Drug Can Protect High-Risk Patients, NY Times March 17, 2017
48. Krumholtz, Harlan, An End to Heart Disease? Not Quite, NY Times March 20, 2017.
49. Roberts, Siobhan, How to Think Like an Epidemiologist, NY Times, August 4, 2020.
50. Tierney, John, And Behind Door No. 1, a Fatal Flaw, NY Times, April 8, 2000
51. Tierney, John, Monkeys, Candy, and Cognitive Dissonance, NY Times, January 27, 2010.
52. Berger, James, and Donald Berry, Statistical Analysis and the Illusion of Objectivity, American Scientist 76, 1988, 159–165.
53. DeVeaux, Michele, Kane, Michael, and Daniel Zeiterman, Clinical Trial Design Using a Stopped Negative Binomial Distribution, Stat. Interface 11, 2018, 699–707
54. Effect of Colonoscopy Screening, New England Journal of Medicine 287, 2022, 1547–1556.
55. Abramson, John, The Price of Colonoscopy, NY Times, February 22, 2011

56. Takacs, Lajos, The Problem of Coincidences, Archive for History of Exact Sciences 21, 1980, 229–244
57. Ross, Ken, A Mathematician in the Ballpark, A Plume book, 2007.
58. Sprenger, Jan, and Naftali Weinberger, Simpson's Paradox, Stanford Encyclopedia of Philosophy, 2021.
59. Simpson, Edward, The Interpretation of Interaction in Contingency Tables, J. Royal Statistical Society (B) 13, 1951, 238–241
60. Berkson's Paradox, Wikipedia
61. Berkson, Joseph, Limitations of the Four-fold Table Analysis to Hospital Data, Biometrics Bulletin 2, 1946, 47–53.
62. Parrondo, Juan, and Luis Dinis, Brownian motion and Gambling: From Ratchets to Paradoxical Games, Contemporary Physics 45, 2004, 147–157
63. Maltese, George, A Simple Proof of the Fundamental Theorem of Markov Chains, American. Mathematical Monthly 93, 1986, 629–630.
64. Blakeslee, Sandra, Paradox in Game Theory: Losing Strategy That Wins, NY Times, January 25, 2000.
65. Sharat, Shannon's Demon, Parrondo's Paradox, Sharat's Newsletter, 2014.
66. Handley, James, Jumping to Coincidencess, The American Statistician 46, 1192, 197–201.
67. Carey, Benedict, Journal's Paper on ESP Expected to Prompt Outrage, NY Times, Jan 5, 2011.
68. Schneps, Leila, and Coralie Colmez, Justice Flunks Math, NY Times, March 26, 2013
69. Richards, Carl, Beware of Market Predictors, NY Times, March 26, 2012
70. Lehrer, Jonah, The Truth Wears Off, The New Yorker, December 13, 2010
71. Bird, Union Fights Fire Department Cuts, NY Times, Dec. 22, 1972
72. Base Rate Fallacy, Wikipedia
73. Bar-Hillel, Maya, and Ruma Falk, Some Teasers Concerning Conditional Probabilities, Cognition 11, 1982, 109–122.
74. Haddon, Mark, The Curious Incident of the Dog in the Night-time, Vintage Books, 2003
75. Samworth, Richard, Stein's Paradox, Eureka 62, 2012, 38–41
76. Fire Department Deployment Analysis, Elsevier, 1979

SPRINGER NATURE

GPSR Compliance

The European Union's (EU) General Product Safety Regulation (GPSR) is a set of rules that requires consumer products to be safe and our obligations to ensure this.

If you have any concerns about our products, you can contact us on ProductSafety@springernature.com

In case Publisher is established outside the EU, the EU authorized representative is:

Springer Nature Customer Service Center GmbH
Europaplatz 3
69115 Heidelberg, Germany

The manufacturer's authorised representative in the EU is Springer Nature Customer Service Centre GmbH, Europaplatz 3, 69115 Heidelberg, Germany. If you have any concerns regarding our products, please contact ProductSafety@springernature.com

Printed and bound by CPI Group (UK) Ltd, Croydon, CR0 4YY
01/10/2026
02225982-0003